STRANGER THAN YOU CAN IMAGINE

BY

Dr. Donald P. Coverdell

Other fine books by Dr. Coverdell include: *The Mystery Clouds, What Manner of Man, Signs and Wonders, Terror in the Air, Texas Hold'em,* **and** *Precious and Very Great Promises.*

Copyright © Dr. Donald P. Coverdell

All rights reserved, including the right to reproduce this book or portions thereof in any form whatsoever except as provided by U.S. Copyright law.

ISBN: 978-1-4357-4541-4

TABLE OF CONTENTS

FORWARD 5

(1) Astonishing Feats
The Will to Survive 8
The Just Out of Reach Treasure 10
The Man Who Lit the World 16
The Stone Giants of Easter Island 22

(2) UFO's, Aliens, & Moon Mysteries
Strangers in the Sky 25
Alien Encounters 37
Ancient Astronauts 44
Moon Mysteries 50

(3) Strange Unexplained Phenomenon
Spontaneous Human Combustion 62
Mysterious Vanishings 66
Psychic Dreams 77

(4) Bizarre Human Oddities
Giants in the Land 89
Dwarfs & the Littlest People 93
Idiot Savants & Mental Marvels 95
Natures Footprints 99

(5) Fascinating Creatures & Botanical Marvels
Strange Things & Living Creatures 108
The Strange Garden Of Nature 121

(6) The Paranormal & The Supernatural
The Near Death Experience 127
Visions of Heaven & Glimpses of Hell 133
Angelic Encounters 139

(7) Mysteries From The Bible
Noah's Ark: Fact or Fiction? 150
Secrets from the Caves 158
Two Men Who Never Died 163
The Secrets of Enoch 170
Mathematical Proof of the Bible's Authenticity 177
(8) The Inexplicably Bizarre
Lives Saved by Animals 185
The Forgotten Genius 192
A Time Warp? 196
Two Men Who Could Fly 197
The Creation of a Life Form? 201
Living Fish in the Pharynx 204

Bibliography And Acknowledgments 206

FORWARD:

Stranger Than You Can Imagine

The span of subject matter in this book is enormous, and the range covers a spectrum. This work is a collection of astonishing stories, bizarre human oddities, incredible events, inexplicable happenings, the paranormal and supernatural, the will to survive, astonishing feats, botanical marvels, extraordinary people, wonders of nature, encounters with angels, UFO's, and alien abductions, fascinating creatures, the Near Death Experience, psychic dreams, and strange unexplained phenomenon. Each chapter can be read independently. The research took more than five years to complete, and is quite compelling.

Do UFO's exist, and have aliens abducted people? Has the United States Government recovered a crashed UFO and alien bodies? Is there a government cover-up, and what really happened at Roswell, New Mexico in 1947? Are there structures on the moon that NASA doesn't want you to see? Can certain people travel out of their bodies and look back and see their physical forms? Do extraordinary paranormal capacities like psychic dreams await discovery in the human brain? Why do certain people suddenly burst into flames for no apparent reason? How can you explain the sudden and inexplicable vanishing of people in plain view of others? What is the truth behind the Philadelphia experiment? What do the Dead Sea Scrolls reveal about man's true nature? Have so-called fallen angels corrupted the scientific and political minds of the planet? Did Giants once walk the Earth as the Bible claims? Who carved the 30 foot tall, 90 ton, stone giants on Easter Island, one of the most isolated spots in the Pacific, and how were they moved more than 10 miles and erected on platforms? Is the Bermuda Triangle an inter-dimensional doorway to another world? Do we have Guardian Angels? Are they with us now? Did you know that many different kinds of animals including dogs, a cat, birds, horses, and even a snake has been credited with the saving of a human life? That one man is responsible for electricity, as we know it today? And how is it possible to know the location of a vast treasure, and yet not be able to recover it?

Here in one scrupulously researched book, these and hundreds of other questions from the world of the unknown, the paranormal, and the extrasensory, are examined in authoritative, compelling, and fully comprehensible terms. You will read the first hand accounts of a number of psychic dreams that actually came true! You will meet people who claim to have left their physical bodies during a Near Death Experience! You will see startling new evidence of life after death! We will discuss studies done on over 1000 sets of identical twins as a criterion of the relative powers of nature and nurture. Hear the stories of people who had the will to survive unspeakable horrors. Learn the unique attributes of Idiot Savants, and mental marvels! Read about the man who could levitate in front of hundreds of witnesses including the Pope! Is there mathematical proof of the Bible's authenticity? This unique book presents the facts and figures, theories and case histories, in terms that will allow you, the reader, to weigh the evidence and decide for yourself, the reality and relevance of the information that is presented in an enjoyable and entertaining format.

1
Astonishing Feats

The Will To Survive

When the Central America left Havana Tuesday, September 8, 1857, she steamed out to sea with some five hundred passengers and a crew of one hundred.

The storm which destroyed the Central America had commenced on the 9th of September, off the shores of Florida, and between that day and the following Sunday it swept the entire coast from Cape Canaveral to Cape Hatteras. The Central America was therefore literally midway in her route when she met the gathering tempest.

The gale increased into a full-blown hurricane, the raging wind ripped the canvas to fluttering tatters. With canvas and rudder and faltering steam engines, they fought the terrible wind and sea.

Saturday, the 12th, dawned upon the passengers still engaged in bailing, but all was of no avail, as the water was increasing rapidly. By noon the Central America was lying well over on her side, with her portholes in the water. The ship steadily lowered into the water.

Captain Herndon ordered the ship's signal gun be fired. The Captain of the brig Marine, which was close by, saw the flares and ordered his ship swung toward the sinking Central America. With one of its two masts snapped off in the storm, she sailed toward them now, with only the canvas of her remaining foremast.

Now woman and children were being lowered into the sinking ships three lifeboats. There was no room left for any of the men. But maybe the woman and children could be saved.

It took two full hours to get the lifeboats along side the brig Marine. The lifeboats were tossed about so violently that the only way of getting out was to watch for a fortunate opportunity and seize hold of the brig's rigging and ropes on the side. But somehow, all the woman and children were now safe on board the brig Marine.

Astonishing Feats

Captain Herndon ordered parts of the upper deck cut loose so that these parts would float free when the ship went under, for the use of the men who would be struggling in the raging water.

By now it was dark, and hundreds of men stood waiting. Some with life preservers and some with planks or wooden fragments of the ship, such as hatch covers and wooden doors, but many had nothing.

The vessel gave three lurches, with some of the men jumping off at each lurch. But the great mass remained on deck. The next instant the ship disappeared under the waves, leaving some 500 men floating in darkness in the storm tossed sea. Many never came to the surface.

Mr. Tice drifted all night on a 10-foot plank. Sunday morning dawned - the storm was now over - the sun now shone brightly, and the heat became almost intolerable. Fortunately for him, his spirit did not fail him for he was determined to struggle for his life. Nightfall came, and alone and exhausted, his plank still in his possession, the long night passed. Monday came and went with agonizing thirst and despair.

On Tuesday, he saw a dark looking object at some distance. He made his way toward it. To his great joy, he discovered it was one of the wooden lifeboats of the Central America. The boat was half filled with water, and contained oars, a pan, a pail, and three old coats. With the pail he bailed out the seawater, he then fastened one of the coats to an oar as a signal. These things accomplished, Tice sat down and rested. His thirst became more and more intolerable, but he resisted the temptation of drinking seawater. Exhausted, he fell asleep, which amid troubled dreams, continued until morning.

Wednesday, long and dreary, followed; his raging thirst and hunger, a burning tropical sun, and for the long twenty four hour days, there was no cessation of agony.

Thursday came, and he noticed something in the distance. A portion of the hurricane deck from the steamer, upon which two men, Alexander Grant, and George Dawson, who had been nearly five days floating upon the open sea.

Tice learned from his fellow sufferers that they were the only surviving ones out of twelve men who had taken refuge on the deck. Grant said it was impossible to sit upright, as the raft would not support them. They were therefore obliged to lie down with their feet in the water, their heads elevated, and their bodies secured by holding on to the ropes with their hands. As may be imagined, their situation was dreadful in the extreme, as the sea constantly washed over them, and one by one, the poor souls slid off into the sea, until only Grant and Dawson remained on the raft. Now the three of them were in the lifeboat.

Friday and Saturday passed, and there was no prospect of relief. The desire for food had ceased, but fierce thirst consumed them. A week had now passed since they were consigned to the deep by the sinking of the steamer. Sunday came and went with no chance of rescue.

Monday's daylight finally dawned, but showed nothing to encourage their hope.

Late Tuesday - a brig appeared a few miles away. It was coming toward them. Suddenly the strained eyes of the sufferers saw the craft unfurl and set the topsail, and then the bow of the vessel was turned upon them. They came near the vessel's side; a line was thrown and caught; and in a few moments, by the aid of truehearted sailors, lines were put around their now totally exhausted forms, and one by one, they were raised upon the deck of the rescuing vessel. They were saved! Saved, after enduring suffering unparalleled in the history of shipwrecks. For nearly nine days, they suffered from hunger, thirst, the sun and the sea. Suffering, which seemed incredible for human nature to bear, and yet continue to live. These three indeed had a will to survive!

The Just Out of Reach Treasure

For nearly 200 years, singly and in syndicates, men have been boring and tunneling on a small island off Nova Scotia in search of a fabulous treasure that many believe lays buried deep beneath the surface.

In recent years the use of electronic devices for locating buried treasure has seemingly confirmed what dozens of frustrated treasure seekers already suspected – that the gold of Oak Island is buried more than one hundred feet deep and has thus far defied all efforts to recover it.

For the treasure of Oak Island, if in fact it is buried there, was so cunningly concealed that it has defied the best efforts of all comers.

Nearly 2 million dollars has been poured into the recovery effort, but no recovery has been made. For whoever dug the so called "money pit" on Oak Island was a brilliant engineer, who harnessed the sea to keep people out.

Astonishing Feats

Oak Island is a tiny rock knoll that sticks up from the cool blue waters of Mahone Bay, in southern Nova Scotia.

Jack Smith and his companions were not treasure hunters at all when they rowed to the island for the first time. But when they noticed the Huge oak tree, several hundred feet inland, with a sawed off limb that showed signs of having once supported a block and tackle. Their curiosity was aroused.

It was Tony Vaugnn who called the attention of his companions to the worn place on the limb of the Huge oak tree and to the depression in the soil, some ten or twelve feet in diameter, directly beneath the same limb. Had the tree been used to lower some very heavy object into a hole? Perhaps this might be a treasure? The three young men could arrive at no other conclusion.

Oak Island had been used as a haven for pirates for many years. The pirates could anchor in one of the convenient deep coves on the north end, post a lookout, and trade with the Nova Scotians for the goods they needed. Teach, Morgan, Bonny, and other notorious pirates had all camped out on this island at one time or another.

The three boys started digging. They found a 13-foot wide circular shaft dug through clay, with thick oak platforms at 10, 20, and 30 feet.

Lacking adequate tools, Smith told Dr. John Lynds of their strange find. That was a fateful day for Dr. Lynds when he first stood there beside that shaft; for in the ensuing years he spent his entire fortune probing its depths for the treasure he was so confident was there.

The diggers had broken through eight oak platforms, one every 10 foot. The last three of which were sealed with ship's putty and coconut fiber. Just under this last platform, a flat stone tablet, which bore unintelligible hieroglyphics, was found. This stone became known as the "cipher stone." Many scholars studied the strange stone, but none had the faintest idea of the message it bore.

At both ninety and hundred feet, they found another and more massive layer of planks and ship's putty.

Dr. Lynds ordered a large drill to be driven through this latest impediment. The instrument slowly made its way thorough the hardwood planks and brittle putty until it appeared to drop into some sort of subterranean room. Lynds was sure he had found a room filled with treasure!

The next day they discovered that the shaft was filled with 60 feet of water. In desperation they brought pumping equipment, which worked round the clock with no effect.

Lynds had his men sink another shaft, close to and parallel with the original. At 110 feet they started burrowing toward what they thought was a room filled with treasure, when all of a sudden, water burst into the shaft - and filled it to the same level as the original hole. Three of the workman drowned. The dazed and battered survivors now had two holes filled with water. The once wealthy Dr. Lynds was now practically penniless, and the work stopped.

Some years later, Dr. Lynds tried again, with the help of a syndicate he had put together.

At 110-foot level, they struck what seemed to be a layer of solid stone. They used a big auger, slowly turned by hand, to dig through the layer of what was found to be hard plaster. The auger then went into several feet of empty space before it touched wood. Four inches of wood, and the workman excitedly announced that they could feel the end of the bit turning in some sort of loose metal! As they pulled the bit up, there, twisted about the screw tip on the drill and jammed into its pod, were three heavy gold links, unquestionably part of a chain! But their work weakened and undermined the shaft itself and caused the bottom to collapse into what was thought to be a vast cavern - carrying down the gold, and dispersing the contents, and the shaft filled with seawater once more.

The syndicate did discover why the shaft flooded to a level that rose and fell with the tide. A man made tunnel, 111 feet down, connected with the sea. The tunnel had been dug from the inlet to the spot where the treasure room seemed to be located, some 400 feet. The engineers reasoned that after the treasure had been stowed away, a thick layer of coconut matting had been placed over the entrance to the tunnel. Achieving a watery trap for anyone who did not know where to find and plug the entrances to the tunnels before trying to enter the treasure chamber itself.

They dug into the beach and at a depth of three feet found a two-inch layer of the same coconut fiber that had been discovered in the money pit. Beneath this fiber came a four or five-inch layer of eelgrass or kelp, and beneath this had been placed a quantity of flat stones. For 150 feet along the beach between high and low water marks the spongy construction continued. Five box drains, strongly built of flat stones at a depth of five feet, led from it to a funnel-shaped sump set just above high water mark.

So well built and protected were the drains that on uncovering them it was found that no earth had sifted through to obstruct the water flowing along it. From the sump the water passed along a tunnel for a distance of 400 feet, sinking steadily until it reached the money pit somewhere below the 90-foot level.

The work involved in sinking the money pit was impressive, but the construction of the flood-tunnel system was unquestionably the work of an engineering genius. As the tide rose in the bay, water was soaked up by the coconut fiber sponge and channeled through the drains and tunnel into the pit. As long as the pit remained undisturbed, the pressure of earth in the shaft held the water back. As earth was removed, the pressure lessened. Then, when the treasure was almost reached, the water broke through from below and flooded the shaft.

These discoveries strengthened the conviction of the treasure seekers that the money pit must conceal an immense hoard. Why else would anyone go to such trouble to protect it?

The tunnel was dynamited and blocked, but still the pit filled uncontrollably, with water from the sea.

In 1865 operations were taken over by a group known as the Halifax Syndicate. They also built a dam at Smith's Cove, which was destroyed by the tide. But they managed to locate the point at which another flood tunnel entered the money pit. This occurred at 110 feet, 10 feet below the level on which the chests were thought to have rested originally. The mouth of the tunnel was about four feet high by two feet wide, lined with round beach stones. The skill of those who had built it was awe-inspiring.

With the realization of the unique system of protecting what ever might be placed in the money pit, how had those who buried the treasure intended to recover it? Some system of floodgates, which could be closed when required, must have been introduced.

In 1874, still another group spent a hundred thousand dollars pumping water from the sea.

Then in 1893 Frederick Blair decided that it was easier to drill through water than pump it out. If the drills brought up any proof that there was treasure down there, Mr. Blair would proceed. The drill brought up three very interesting bits of evidence: First, it showed unmistakable signs of having drilled through gold, second, it showed that there were layers of wooden casks or boxes filled with gold to a depth of at least thirty feet, and these containers were in a room made of heavy lumber and coated with crude cement about twenty inches thick; third, one of the drills brought up scraps of parchment, which bore letters.

After four years of hard work, the group headed by Blair had to admit defeat. Blair struggled on alone until 1903, when poverty overtook him.

In 1909 came Captain Harry Bowdoin, a New York engineer with rich and powerful friends. Bowdoin cleared out the Money Pit to a depth of 113 feet and put down a core drill. At 149 feet they drilled through what appeared to be a layer of cement. It was suggested that this could be the roof of a treasure chamber. But below this layer the drill brought up only yellow clay. But in the resulting patchwork of shafts, the exact location of the money pit was lost.

In 1931 William Chappell, who had been in the 1897 syndicate with Blair, returned to Oak Island, and sank a shaft where he thought the money pit was. He uncovered a pick, a miner's seal-oil lamp, and an ax head thought to be 250 years old! He came to believe that the treasure had dropped further down into some natural cavity in the rock as a result of all the activity in the money pit. Greater understanding of the geological formation of the Island was later to show that in fact the underlying limestone contained many cavities and sinkholes.

In 1933 came a British Colombian named Thomas M. Nixon, who believed that the Incas had deposited a treasure on Oak Island. He proposed to enclose the entire money pit area within a circle of interlocking steel pipe. This was indeed a good plan. However, he did no more than sink a few boreholes and depart.

Then came Gilbert D. Hedden who made a significant discovery in the tangled undergrowth on the edge of the south shore. A triangle of beach stones had been arranged on the ground with a curved line enclosing the base to give it the appearance of a sextant. The sides of the triangle were 10 feet long, and an arrow made of stones slanted across the base line to the apex. This arrow pointed directly at the money pit.

Hedden did not stumble on this sign by chance. He was led to it by following directions in the book *Captain Kidd and his Skeleton Island*, written in 1935 by Harold Wilkins. This book included a chart of an Island that resembled Oak Island. Hedden was convinced that the treasure in the money pit was Captain Kidd's.

Wilkins explained that he had drawn the map in the book after being allowed to look at some authentic Kidd charts in the possession of

Hubert Palmer, a collector of antiques. All the charts depicted the same curving island and bore the date 1669. The charts, which are now in Canada, are considered to be genuine.

A second tunnel, at 150 feet was discovered in 1942. By Edwin H. Hamilton, a machine engineer. He drilled down to 180 feet, and made two important discoveries. He found that the second flood tunnel entered the money pit at 150 feet and from the same side as the first one, which had been constructed at 110 feet. Both tunnels therefore came from Smith's Cove, one above the other.

All of the digging, drilling, and pumping have so confused the area that the exact location of the original shaft is no longer known.

There, for the moment, the matter rests. But the question remains - who buried what, and why?

Steven Sora, author of *The Lost Treasure of the Knights Templar* believes that this vast treasure was brought to Oak Island and sealed away. The Guise family would know the existence of the vault, but the location would be a secret passed only within the family Sinclair. Subsequent wars and religious strife took a serious toll on the family. Along the way, Sinclairs would die suddenly, in war, in prison, and possibly of natural causes. At some point the secret location was not passed down to the next generation and was lost. Sora makes a strong argument that engineering skills known to the Templars were utilized in digging the deep shaft, the long water tunnels, and the false beach and concealed drains. This book is a must read for any serious Oak Island treasure hunter.

The other hypothesis, put forth by Frank Edwards author of *Oak Island's Untouchable Treasure* deals with the known fact that in the middle of the eighteenth century, when France was wrangling with England over claims in the New World, the French government decided to make Fort Louisburg into an impregnable base of operations. Ten million dollars in gold was sent for that purpose. It is a matter of record that the corrupt Governor spent only a small portion of that sum in improving the fortifications, and there is no indication that the Governor sent any of the gold back.

We know there is a cement and wood chamber, which contains gold. It was concealed by lengthy and extremely clever feats of engineering, performed at considerable cost. But what is buried there and by whom, and why, is still an unsolved mystery.

The Man Who Lit The World

Margaret Cheney, a science writer, and author of the *Tesla: Man Out Of Time,* writes: "He was flamboyant, eccentric, almost supernaturally gifted, had he been born today he would still be ahead of his time. Called a madman by some, a genius by others, and an enigma by nearly everyone, Nikola Tesla was perhaps the greatest inventor the world has ever known.

He was a trailblazer who created astonishing, world-transforming devices, often without theoretical precedent. It was Tesla who harnessed the alternating electrical current we use today. Tesla actually invented radio, Tesla invented fluorescent lighting and the incredible blade-less turbine. He introduced us to the fundamentals of robots, computer and missile science, which continue to create and transform the future."

The International Tesla Society, a non-profit educational and scientific organization in Colorado Springs writes: "Tesla's inventions are monumental. During his life (1856-1943), he registered over 700 patents worldwide. They include radio, the alternating current motor, high frequency and high voltage circuitry, logic circuits used in today's electronics, electro-therapeutics and other areas of technology as we know it today."

Often referred to as a man out of time, Nikola Tesla was born at precisely midnight between July 9 and 10, 1856, in the village of Smiljan, province of Lika, Croatia. Tesla was of Serbian heritage, his father an Orthodox Serbian Priest and his mother from a long line of inventors.

Throughout his childhood, Tesla was tormented by extraordinary mental images, which he inadvertently created in his mind. Later, he learned to control these thoughts and was able to visualize his experiments without ever having to construct them. In fact, when it came to his mechanical devices, he was able to visualize even the wear and tear on the equipment. He was, therefore, able to "test" his inventions in his mind - long before putting them on paper or making a model.

Astonishing Feats

He wrote, "It should be possible to project on a screen the image of any object one conceives and make it visible. Such an advance would revolutionize all human relations. I am convinced that this wonder can and will be accomplished in time to come; I may add that I have devoted much thought to the solution of the problem."

During his adult life, Tesla was quite a celebrity. He regularly entertained such notables as Mark Twain in his New York lab.

Chaunceey McGovern, author of *The New Wizard Of The West* writes: "Fancy yourself seated in a large, well-lighted room, with mountains of curious-looking machinery on all sides. A tall, thin young man walks up to you, and by merely snapping his fingers creates instantaneously a ball of leaping red flame, and holds it calmly in his hands. As you gaze you are surprised to see it does not burn his fingers. He lets it fall upon his clothing, on his hair, into your lap, and, finally, puts the ball of flame into a wooden box. You are amazed to see that nowhere does the flame leave the slightest trace, and you rub your eyes to make sure you are not asleep."

If McGovern was baffled by Tesla's fireball, he was at least not alone. None of his contemporaries could explain how Tesla produced this oft-repeated effect, and no one can explain it yet today!

Befriended by George Westinghouse, Tesla was able to see the implementation of his AC electrical system despite stiff opposition by Thomas Edison and his associates.

At the time, Edison had opened the Edison Machine Works on Goerck Street and the Edison Electric Light Company at 65 Fifth Avenue. His direct current generating station on Pearl Street was serving a few hundred people in the Wall Street and East River area, and he had a big research laboratory at Menlo Park, New Jersey. But the big problem with direct current was that it required a generating station every few miles, and the primitive Edison dynamos broke down often, and did not work efficiently.

Meanwhile, Tesla, whose engineering reputation was beginning to be favorably known, was approached by a group of investors, and the Tesla Electric Light Company was formed, with headquarters at Rahway, New Jersey.

Working in his first laboratory he developed the Tesla arc lamp. This system was put to work as streetlights in Rahway.

Because it was all there in his mind he needed only a few months to design and patent the complete poly-phase AC system, which Edison saw as the death knell for his own direct-current system.

George Westinghouse saw the revolution coming and entered into a contract with Tesla giving him a royalty of $2.50 per horsepower of electricity sold, which soon made Tesla a millionaire.

Westinghouse was doing well in his drive to put the country on an alternating-current system. But there was one big problem. Nikola Tesla's patent royalties already more than 12 million dollars, was holding back his expansion plans.

Reluctantly he called on the inventor to give up his contract. Westinghouse told him: "I believe your alternating current system is the greatest discovery in the field of electricity. It was my efforts to make it available to the world. But this contract is really holding me back."

Tesla looked at Westinghouse and replied: "I will tear up the contracts, and you will no longer have any troubles from my royalties. Is that sufficient?"

By destroying the contracts, Tesla not only relinquished his claim to millions of dollars in already earned royalties but to all that would have accrued in the future, which would have been countless billions!

Tesla, deep in his research, was scarcely aware of the millions of dollars being made on his inventions by Westinghouse. He was immersed in a whole new world of electrical phenomena, while Westinghouse built and sold Tesla motors and generators.

As long as the world left him alone in his laboratory to pursue his love affair with electricity, Tesla was the happiest man alive. He had introduced the power system that was to revolutionize industry and light the world, bringing light to even the most remote homes.

Tesla next revealed the fluorescent light. But it did not find it's way to commercial markets until fifty years later!

Roland J. Morin, the chief engineer of Sylvania GTE International wrote: "I am sure that Tesla's demonstration of these light sources at the Chicago World's Fair in 1893 stimulated D. McFarlan Moore to develop and announce commercial realization of the fluorescent lamp."

One of Tesla's most grandiose conceptions, the "terrestrial night light" was a way of lighting the whole Earth and its surrounding atmosphere, as though it were but a single illumination. He theorized that the gases in the atmosphere at high altitudes were in the same condition as the air in his partially evacuated radio tubes and hence would serve as excellent conductors of high-frequency currents.

Astonishing Feats

Tesla also talked of a plan for using both earth and upper air as conductors of electricity and the stratum of air between as an insulator. This combination would form a kind of gigantic condenser, a means of storing and discharging electricity. If the Earth were electrically excited, the upper air would be charged by induction. A current flowing both in the ground and in the upper air would create a luminous upper stratum that would light the world.

Then, suddenly disaster struck. At 2:30 in the morning of March 13, 1895, his laboratory caught fire and burnt to the ground. Stunned, sickened, he turned away from the ruins in the cold early morning and wandered through the streets in a trance, paying no attention to where he was or to the passing of time.

Newspapers all over the world reported the tragedy. The headlines read: **"Work of a Lifetime Gone." "Fruits of Genius Swept Away."** Charles A. Dana of the New York Sun wrote: "The destruction of Nikola Tesla's workshop, with all its wonderful contents, is something more than a private calamity. It is a misfortune to the whole world. It is not in any degree an exaggeration to say that the men living at this time who are more important to the human race than this gentleman can be counted on the fingers of one hand: perhaps on the thumb of one hand."

Tesla was not downcast for long, however, consoling himself with the fact that his ongoing research was still vivid in his mind and that his loss was merely a setback. He soon was at work in his new laboratory.

One day in 1898 while testing a tiny electromechanical oscillator, he attached it with innocent intent to an iron pillar that went down through the center of his loft building at 46 East Houston Street, to the sandy floor of the basement.

Flipping on the switch, he settled into a straight-backed chair to watch and make notes of everything that happened. Such machines always fascinated him because, as the tempo built higher and higher, they would establish resonance with first one object in his workshop and then another. For example, a piece of equipment or furniture would suddenly begin to shimmy and dance.

What Tesla was unaware of on this occasion was that vibrations from the oscillator were being carried through the substructure of Manhattan in all directions. Buildings began to shake, windows shattered, and people ran out into the streets in fear of this man made earthquake!

Tesla had just begun to sense an ominous vibration in the floor and walls of his building. Knowing that he must quickly put a stop to it, he seized a sledgehammer and smashed the little oscillator in a single blow.

In later years, he talked of having the knowledge to build an "earthquake machine" that could topple whole cities. But he insisted that it would be too dangerous to build one. However, the Army wanted to contract for such a machine, but Tesla refused, saying he did not want to be responsible for any "mass destruction."

Nothing was to be developed from this concept, but Tesla put the fear of God into New Yorkers when he talked of the awesome potential of mechanical resonance. He could walk over to the Empire State Building, he told reporters, "and reduce it to a tangled mass of wreckage in a very short time." The mechanism, he said, "Would be a tiny oscillator you could slip into your pocket."

Later, betrayed by the magnates of his time including J.P. Morgan, he continued to invent many devices for the betterment of life as we have it today. He was much chagrined, however, by the fact that Marconi received credit for the invention of the radio. It was not until after his death in 1943 that the US Supreme Court overturned the patents previously awarded to Marconi for the radio and gave them back to Tesla. However, since he had died just three months earlier, there was no one available to help correct history. Today, many school children are still taught that Marconi invented the radio.

The demonstration of the world's first radio-controlled robot boat by Tesla in 1898 was the common ancestor of modern guided weapons and vehicles, of automated industry, and of robots. The world would not be ready for these ideas for many years to come.

In Colorado Springs, Tesla intended to develop a worldwide wireless system for communications well ahead of the ambitious Marconi. The land he was using, in the shadow of Pike's Peak, was 6,000 feet above sea level. The air was clear, dry, and crackling with static electricity. He disclosed that he planned to send a wireless message from Pike's Peak to Paris in time for the Paris Exposition of 1900.

A fence surrounded the weird structure that began to rise from the prairie floor looking like a large square barn. Extruding from an open section of the roof was a tower that reached eighty feet above the ground. From this a metal mast soared another 122 feet. Poised upon its tip was a copper ball three feet in diameter.

Astonishing Feats

Machinery was moved in an assembled as quickly as it arrived on the construction site.

The message he would transmit to the Paris Exposition of 1900 was - "my greetings to the crazy French!"

Tesla wrote, "that by use of the magnifying transmitter, I have passed a current around the globe."

Another of Tesla's claimed discoveries at Colorado Springs came late one night as he was working at his powerful and sensitive radio receiver. Tesla claimed to have received a message of extraterrestrial origin! Tesla wrote that the signals came from "Intelligent beings on a neighboring planet."

Tesla wanted to reply to these "messages" from outer space. He felt that he was on the threshold of a new revolutionary technology. He began to work on the instruments he would need for such an undertaking. But he soon ran out of money, and the project came to a halt.

Tesla's greatest scientific gift to the world, one that has never been brought forth commercially due to financial considerations, was the transmission of electricity without wires!

With this system, all that would be required to receive "free" electricity for your home or work place would be an antenna erected into the air. He demonstrated this principle while in Colorado Springs in 1899.

The electricity would be "transmitted" through the air, much the same way as present day radio waves. However, Westinghouse, Edison, and the US government said such a thing, as free energy would bankrupt the economy.

With fresh funds from J.P. Morgan, Tesla had built just such a transmitting tower behind his new research facilities at Wardenclyffe. The fifty-five ton, sixty eight-foot dome frame structure looked much like a water tower. The dome was to be covered with copper plates to form an insulated ball. The bottom was firmly anchored into the earth. In the top of the strange tower were large coils of wire, which acted as giant transformers. But the tower was not yet finished, and Morgan would not make any more advances, and Tesla was now broke.

The humiliating news of Tesla's financial distress followed by his loss of Wardenclyffe was further advertised in March 1916 when he was summoned to court in New York for failure to pay $935 to the city in personal taxes. The land on which Wardenclyffe stood was taken from him and sold to a New York attorney.

Most people today have not even heard the name Tesla. But he probably has had more effect on life, as we know it, than any one else, including men like Einstein and Edison. Without Tesla, the automobile would not be able to run, because every gas powered auto uses a "Tesla Coil" to cause the spark plugs to fire. Every time you turn on a light switch, you can thank Tesla for the AC current which flows through your wires. Many of the marvels of today have their roots, indeed, their primary patents, traceable to one individual - Nikola Tesla, the man who lit the world.

The Stone Giants Of Easter Island

Dutch admiral Jakob Roggeveen had never seen anything like it in his life before - an uncharted island in the mid-South Pacific inhabited by what appeared to be 30 foot tall stone giants. Each one was standing on a large stone platform. They all had long ears, and a red hairpiece.

The date was Easter Sunday 1722, and so Roggeveen named his discovery Easter Island. He then sailed away.

One of the most isolated spots in the Pacific, tiny Easter Island is 2300 miles west of the coast of Chile, to which it has belonged since 1888.

There is evidence that lush vegetation and considerable timber may once have flourished on the island, but today its plant and animal life are sparse.

These massive, stylized figures of volcanic stone are both majestic and disturbing. Their heads are immense, their expressions brooding and disdainful, their ears grotesquely elongated, their chins jutting and powerful. Their arms hang rigidly at the sides of their legless trunks; their hands extend stiffly across their protuberant bellies. So far some 1,000 of them have been found, many weighing 90 tons or more and standing as high as 32 feet. Unfinished statues more than twice this size have been found.

What captures the imagination even more than the forbidding magnificence of these stone giants is the fact that they exist at all in such an unlikely spot. What happened to the men who carved them? How were the statues moved from their quarries and raised on the stone altars, and where is the timber that must have been used to accomplish this task? Above all, why are there so many statues, why are they so large and what was the purpose they served?

Astonishing Feats

The giant statues carved in the dormant crater of Rano Raraku, had been lowered down its slope, and somehow maneuvered into upright positions on the platforms.

In side the crater were some 400 uncompleted statues. Some that was almost ready for transportation. They had been left as though the craftsmen had intended to return some day but never did. Down the road leading from the crater were dozens of completed statues. They lay scattered every 50 yards for as far as the eye could see. One unfinished giant was 66 feet tall, and weighed 50 tons.

Some of the statues were 10 miles or more from the crater, and experts have still not worked out how they were transported, or erected.

Theories that logs were used as rollers were discounted, after tests showed that the Easter Island soil could not support trees of the size required for such an exercise. Another possibility considered was that vines woven into ropes were used to haul them. But this, too, was ruled out when it was proved that vine ropes could not stand the strain of pulling 30 tons.

Yet in size, number and brooding configuration of face and torso, no colossi can match the stone giants scattered over the grasslands and volcanoes of Easter Island.

Thor Heyerdahl, the Norwegian anthropologist and adventurer who mounted an archeological expedition to Easter Island during the 1950s, theorizes that the first settlers, the Long-ears, came from Peru, perhaps as early as A.D. 300, since the Easter Island colossi bear some resemblance to ancient statues found in South America. A second clue is the profusion of wood carved ideograms found on the island, all still undeciphered, and all inscribed on hardwood tablets known as Rongo-Rongo. The people of ancient Polynesia, Heyerdahl points out, unlike the Peruvians, had no written language. Heyerdahl's most important argument for Peruvian derivation, however, comes from his famous voyage on the raft Kon Tiki in 1947, to prove that Polynesia was originally settled by Indians sailing westward from South America rather than eastward from Asian lands.

The huge, non-mortared basalt blocks that face the platforms are so smoothly and precisely fitted that it would be difficult to pass a knife blade between them. The Stonework here bears a strong resemblance to Inca masonry.

Perhaps the most enigmatic of Easter Island's treasure are in the inscribed wooden tablets called collectively as the "talking boards." Only 26 of them remain in existence. All are covered with minute, beautifully incised characters, largely derived from human and animal forms. Thomas Barthel, a German ethnologist thinks the language is of Polynesian origin.

The Islanders that Roggeveen met were totally isolated, unaware that other people existed. Investigators in all the years since his visit have discovered no trace of the islanders' having any outside contacts.

Easter Island's mysteries have spawned volumes of speculation for more than two and a half centuries. Many Europeans were incredulous that Polynesians - commonly characterized as "mere savages" - could have created the statues or the beautifully constructed stone platforms.

In 1960, the Swiss writer Erich Von Daniken, an ardent believer in Earth visits by extraterrestrial astronauts, said that he felt the Easter Island statues were the work of intelligent beings who owned ultramodern tools, who became stranded on Easter, and were finally rescued.

Heyerdahl and Von Daniken both brushed aside overwhelming evidence that the Easter Islanders were typical Polynesians derived from Asia rather than from the Americas and that their culture, including their statues, grew out of Polynesian culture. Their language was Polynesian, specifically; they spoke an eastern Polynesian dialect related to Hawaiian. DNA extracted from 12 Easter Island skeletons was also shown to be Polynesian. The Islanders grew bananas, taro, sweet potatoes, sugarcane, and paper mulberry. All were typical Polynesian crops mostly of Southeast Asian origin. Their sole domestic animal, the chicken, was also typically Polynesian and ultimately Asian.

This volcanic island, lost in the vastness of the Pacific Ocean, cut off in one of the loneliest spots on earth, managed to create a race of stone giants so big that even our engineers still marvel.

Modern research cannot show how the islanders could have built, or moved, or erected these colossal statues weighing more than 90 tons. But the real mystery is...why?

2
UFO's, Aliens, & Moon Mysteries

Strangers In The Sky

The flying saucer story began on June 24, 1947, when a private pilot, Kenneth Arnold, who was taking part in a search for a missing transport aircraft, saw nine disk-shaped objects flying over Mount Rainier, Washington.

He described the objects as looking like "two saucers, one inverted on top of the other."

Arnold's veracity and professional reputation were beyond question, and the authorities accepted his account of the incident. Newspapers picked up the story and labeled the objects "flying saucers."

What followed was a phenomenon in its own right. In the next five years or so, thousands upon thousands of sightings of unidentified flying objects would be seen over North America. The sightings came in waves, periods of relative quiet ending with floods of reports in a single month.

One startling occurrence took place in the southwestern desert several weeks after the Arnold sighting. It was the first report of a crashed UFO.

The Roswell Daily Record, on Tuesday, July 8, 1947, ran the following headline in bold print: **"RAAF Captures Flying Saucer On Ranch In Roswell Region."**
A sub-headline read: **"No Details Of Flying Disk are revealed."** And yet a third sub-headline read:
Roswell Hardware Man and Wife Report Disk Seen." The story read as follows:

> The intelligence office of the 509th Bombardment group at Roswell Army Air Field announced at noon today, that the field has come into possession of a flying saucer.
>
> According to the information released by Maj. J. A. Marcell, intelligence officer. The disk was recovered on a ranch

in the Roswell vicinity, after an unidentified rancher had notified Sheriff Geo. Wilcox, here, that he had found the instrument on his premises.

Major Marcel and a detail from his department went to the ranch and recovered the disk, it was stated.

After the intelligence officer here had inspected the instrument it was flown to "higher headquarters."

The intelligence office stated no details of the saucer's construction or its appearance had been revealed.

Mr. and Mrs. Dan Wilmot apparently were the only persons in Roswell who have seen what they thought was a flying disk.

They were sitting on their porch at 105 South Penn. Last Wednesday night at about ten minutes before ten o'clock when a large glowing object zoomed out of the sky from the southeast, going in a northwesterly direction at a high rate of speed.

Wilmot called Mrs. Wilmot's attention to it and both ran out into the yard to watch.

Wilmot said that it appeared to be about 1,500 feet high. He estimated the speed at 500 miles per hour.

In appearance it looked oval in shape like two inverted saucers, faced mouth to mouth. The entire body glowed as though light were showing through from inside.

Wilmot said he could hear no sound. The object came into view from the southeast and disappeared over the treetops in the general vicinity of six mile hill.

Wilmot, who is one of the most respected and reliable citizens in town, decided that he would go ahead and tell about seeing it. The announcement that the RAAF was in possession of one came only a few minutes after he had decided to release the details of what he had seen."

A second sub headline read: **"Air Force General Says Army Not Doing Experiments."**

The incident was marked by confusion and conflicting accounts. The Air Force first confirmed the story, and then denied it. But something very strange happened that night.

Mac Brazel, a sheep rancher north west of Roswell, said he heard a tremendous explosion that night. The next day he found wreckage scattered over a quarter mile of ground.

Although there were few other details at the time, telephone lines to the base were tied up for days, but the Air Force said nothing more.

Then the cover-up began. The strange disk fragments were taken to eighth Air Force Headquarters in Fort Worth, Texas, Where Brigadier General Roger Ramey went on the radio to call it a mistake. What Mac Brazel had found, said the general was a weather balloon. However, in 1987 documents would surface allegedly showing that a spacecraft and the bodies of four crewmen had been recovered and kept from public view!

In the wake of the Roswell incident, UFO reports surged again. The very next week the first photos purporting to show UFO's in flight were snapped in Phoenix, Arizona. The pictures ran in the Arizona Republic on July 9.

An FBI agent, who asked him for the negatives for evaluation, visited the photographer, William Rhodes. But when Rhodes asked for the negatives back some months later, he was informed by letter that they could not be returned!

The modern history of seeing things in the skies is a voluminous and disturbing record of eyewitness reports, often buttressed with impressive documentation, that have survived the scrutiny of many scientists.

One of the most spectacular recent reports came from a Gemini 4 astronaut, James McDivitt, in June 1965. In orbit, about 90 miles above the earth, McDivitt saw a cylindrical object, apparently with protrusions sticking out of it, which he took to be another spacecraft with antennae. It appeared to be in free drifting flight over the Pacific, somewhat higher the Gemini Capsule. McDivitt took one still photograph and some movie film.

He observed that the object was nearby and moving in a path toward his own spacecraft, but closing in fast. McDivitt and his fellow astronaut, Edward White, were hastily preparing to take evasive action when the UFO disappeared from view.

In a prepared statement, put before the Committee on Science and Astronautics of the House of Representatives in 1968, Prof. James E McDonald, a physicist from the University of Arizona, said, "My own present opinion, based on two years of careful study, is that UFO's are probably extraterrestrial devices engaged in something that might very tentatively be termed surveillance."

In 1978, a panel of scientists (commissioned by the French government) studied UFO sightings in great detail. The panels' originally secret report concluded that something real lay behind the sightings. "A flying machine - whose modes of sustenance and propulsion are beyond our knowledge."

Reports of UFO's have come from no fewer than 139 countries all over the globe, from such tiny nations as Grenada and Kuwait as well as from such major powers as the United States, Russia, England, France and China.

One of the earliest reports of a close look at a passing UFO is that of two eastern Airlines pilots, Clarence S. Chiles and John B. Whitted, on July 24, 1948.

At 2:45 A.M., 20 miles west of Montgomery, Alabama, they saw an aircraft streaking toward them and thought at first that it was a jet fighter. "It was heading southwest, exactly opposite our course. Whatever it was, flashed down towards us with terrific speed. We veered to the left. It veered sharply, too, and passed us about 700 feet to the right. I saw then that it had no wings," said officer Chiles adding to the description: "An intense dark glow came from the side of the ship and ran the entire length of the fuselage - like a blue fluorescent light. The exhaust was a red-orange flame."

Chiles then visited the cabin to check the passengers. Clarence McKelvie was the only one awake. He too had seen a brilliant flash of light pass the window. "It looked like a cigar with a cherry flame going out the back. There was a row of windows. It disappeared very quickly." McKelvie said.

At 9:05 P.M. on June 29, 1954, Capt. James Howard and his crew on board a BOAC Centaurus Stratocruiser observed that there was one big-lighted object with six smaller ones in attendance. The objects, about five miles away, were flying parallel to the Centaurus. These UFO's stayed with the Centaurus for 18 minutes and a distance of 80 miles. After a few minutes of these startling observations, Captain Howard contacted flight personal at Goose Bay. They replied that a fighter pilot was being sent up to investigate.

By this time all 8 members of the crew and 14 of the passengers who were awake were observing the phenomenon, when the smaller UFO's started to disappear. George Allen, the navigating officer who had been watching closely the entire time, later reported, "It looked to me as though they went inside the big one." Finally the big object departed "at tremendous speed."

Captain Howard told the arriving jet interceptor about the gradual disappearance of the UFO fleet and its accompanying "base ship." The captain later wrote in the December 11, 1954, issue of *Everybody's Weekly*: "There is no rational explanation - except on the basis of spaceships and flying saucers. On that basis it must have been some weird form of space ship from another world." He remained convinced that what he saw were solid objects, "maneuverable and controlled intelligently - a sort of base ship linked somehow with those smaller attendant satellites."

Throughout history man has witnessed, and struggled to explain, the presence of bizarre objects in the sky. Objects so strange and foreign to our daily terrestrial mode of thought that it is frequently met by ridicule by persons unacquainted with the facts. Yet the phenomenon persists; it has not faded away. Instead, it has touched on the lives of an increasing number of people around the world.

Just east of New Zealand's South Island, a three member Melbourne television crew was airborne in an Argosy cargo plane, retracing the aerial route between Wellington and Christ Church, a route along which the crews of two aircraft had spotted bright-unidentified lights 10 days before.

At the same time, unexplained radar images were detected by Wellington radar. One image seemed to pace the plane, as witnesses on board watched a flashing light that appeared for a few minutes. During the return flight, a light approached within 10 miles of the cargo plane. This light was described by a TV crewman as having a "brightly lit bottom and transparent sort of sphere on top."

Whatever the nature of the object, it was subjected to remarkably comprehensive human and electronic scrutiny. For it not only was seen by the TV crew and the planes two pilots but it also was tracked by the plane's airborne radar. Most extraordinary of all, the object's elusive presence was documented on color film.

The film revealed an intriguing, if brief, series of images of mysterious flying objects. One sequence showed a bell shaped form that was right on the bottom, as described by a cameraman at the time. A single frame of this sequence showed what seems to be the track of the object as it moved in a classic loop, indicating extremely rapid relative motion between the camera and the object.

The New Zealand sighting marked the first time in history that on the spot tape recordings were made while UFO's were observed, filmed, and simultaneously tracked on radar.

Although it was remarkable, the New Zealand case is unique only in the breadth of its documentation. For in the past thirty years, there have been at least 100,000 reports of mysterious objects in the sky and countless more, perhaps 10 times as many that have gone unreported.

Nocturnal lights are by far the most frequently reported type of sighting. These lights - alone or in groups - hover motionless or dart about the skies in trajectories unlike those of known phenomena. Daylight discs cover daytime sightings such as the one seen by Kenneth Arnold near Mount Ranier in 1947. These disc or saucer shaped UFO's often give off a fluorescent glow. Radar visual sightings cover incidents observed simultaneously by witnesses and radar, as in the case of the sightings over New Zealand in the early morning hours of December 31, 1978.

The second major category, for sightings from a distance of less than 500 feet, was broken down as follows: *Close Encounters of the First Kind* - those in which there is no interaction between the UFO and the environment; *Close Encounters of the Second Kind* - those manifesting some interaction, such as interference with car ignition systems, burns on the ground, and physical effects upon animals or humans; and *Close Encounters of the Third Kind* - those in which UFO occupants from outer space are reportedly seen.

In recent years UFO witnesses have described personal contact with the occupants, and even temporary detainment. These cases are often referred to as *Close Encounters of the Fourth Kind.* (In the next chapter, I will present a number of these cases.)

A vivid example of a CE-1 is the case of Dale Spaur, a deputy sheriff in Portage County, Ohio, which was investigated by the U.S. Air Force's project Blue Book.

Early on the morning of April 17, 1966, Spaur and another deputy stopped at the side of the road to investigate an abandoned car. Suddenly, rising above a nearby woods, Spaur recalls, "I saw this thing as big as a house and bright enough to make your eyes water. It was topped by a dome and gleaming with an intense purplish white light." Spaur and his companion radioed their bizarre report to headquarters and were ordered to give chase.

Across northern Ohio and into Pennsylvania, for more than 70 miles, they chased the object at speeds of up to 105 miles per hour. Midway in the chase, they were joined by a second cruiser, manned by a policeman who had monitored the chase on his radio and then spotted the UFO. The chase ended in Conway, Pennsylvania, where a forth policeman told the others he had been watching the object for the past 10 minutes. Together, the four watched as whatever it was shot straight up into the air and disappeared.

In contrast to the CE-1 cases, *Close Encounters of the Second Kind*, leave a tangible calling card - a measurable effect on animate or inanimate matter. Consider the following CE-II, one of the strangest on record because of the damage done to man made structures. The encounter occurred in the small village of Saladare, in Ethiopia, at 11:30 in the morning, on August 7, 1970, and lasted for about 10 minutes.

A red glowing ball swept over the village, destroying houses, knocking down the stone walls of a bridge, uprooting trees, and melting asphalt and cooking utensils, but strangely enough, not starting fires.

In Scandinavia, in 1972, several mysterious radioactive impressions were discovered in Mans Fjord. In the same year in Rumania, a night watchman reported seeing a strange object descend and later found a perfect circle in the corn field, with a hole six inches wide and eight feet deep at its center. And only a few years earlier, in Valensole, France, a farmer spotted a rugby-ball -like metal object on four legs in a field of lavender. When the UFO zoomed away, it left behind a patch of flattened, scorched vegetation where, the story goes, the farmer was never again able to grow lavender.

Witnesses in CE-II's often report another type of phenomenon - interference with electrical circuitry. In the apparent presence of a UFO, television sets go on the blink, car headlights dim or go out, automobile engines sputter and die. When the UFO disappears, witnesses report, everything works again.

CE-III cases are clearly the strangest of all. They seem preposterous, the realm of science fiction, for *Close Encounters of the Third Kind* involve human like creatures and very often the boarding of UFO's themselves.

Fully one third of CE-III's have involved more than one observer. The multi-witness case that stands out most dramatically occurred in 1959 in Papua New Guinea. The scene was an Anglican mission station, and the principal witness, there were said to be 38 witnesses in all, was Father William Gill, an Australian priest.

The evening of June 26, Gill and others at the mission found them selves watching "this sparkling object," described as a large circular shaped craft with a wide base on four legs. By Gill's account, which was backed by the signatures of 25 of the witnesses, the craft hovered at about 300 feet. Bathed in a blue light that flowed upward, there appeared on top of the craft four human figures. The craft, sometimes accompanied by a pair of smaller objects, was visible off and on for about four hours. The following evening, the UFO reappeared over the mission. Once more, Father Gill and about a dozen onlookers saw the four figures. "Two of the figures seemed to be doing something near the center of the deck, they were occasionally bending over and raising their arms as though adjusting or setting up something not visible. One figure seemed to be standing, looking down at us. I stretched my arm above my head and waved. To our surprise the figure did the same."

Soon, recalled Gill, he and a New Guineau assistant were "waving our arms, and all four seemed to wave back. There seemed to be no doubt that our movements were answered. All the mission boys made audible gasps of either joy or surprise or both."

An "occupant sighting case" was reported on April 24, 1964, in Socorro, New Mexico. State policeman Lonnie Zamora gave up chasing a speeding motorist to chase a UFO instead, when he saw one landing in a field about a mile away.

Zamora reported that he saw a bright, metallic oval, about the size of an upturned car. Standing beside it, he said, "Were two humanoid figures, about the size of 10 year old children." As he called for backup, the figures retreated inside, and the object took off.

On May 22, 1955, an International News Service syndicated report from London by the American journalist Dorthy Kilgallen reported:

> "British scientists and airman, after examining the wreckage of one mysterious flying ship, are convinced these strange aerial objects are not optical illusions or Soviet inventions, but are flying saucers which originate on another planet.
> The source of my information is a British official of cabinet rank who prefers to remain unidentified. "We believe, on the basis of our inquiry thus far, that the saucers were staffed by small men - probably under four feet tall. It's frightening, but there is no denying the flying saucers come from another planet."
> These official quoted scientists as saying a flying ship of this type could not possibly have been constructed on earth. The British government is withholding an official report on the flying

saucer examination at this time, possibly because it does not wish to frighten the public."

One of the most sensational UFO events ever reported by military personal occurred in Randlesham Forest, just outside the perimeter of RAF/USAF Woodbridge, near Ipswich, Suffolk. Charles Halt, U.S. Air Force Deputy Base Commander at Woodbridge filed this official report with the Ministry of Defense on January 13, 1981:

Subject: Unexplained Lights
To: RAF/CC
(1) Early in the morning of 27 Dec 80, two USAF security police patrolmen saw unusual lights outside the back gate at RAF Woodbridge. Thinking an aircraft might have crashed or been forced down, they called for permission to go outside the gate to investigate. The on duty flight chief responded and allowed three patrolmen to proceed on foot. The individuals reported seeing a strange glowing object in the forest. The object was described as being metallic in appearance and triangular in shape, approximately two to three meters across the base and approximately two meters high. It illuminated the entire forest with a white light. The object itself had a pulsing red light on top and a bank of blue lights underneath. The object was hovering or on legs. As the patrolmen approached the object, it maneuvered through the trees and disappeared. At this time the animals on a nearby farm went into a frenzy. The object was briefly sighted approximately an hour later near the back gate.
(2) The next day, three depressions 2 inches deep and 7 inches in diameter were found where the object had been sighted on the ground. The following night the area was checked for radiation. Beta/Gamma readings of 0.1 milliroentgens were recorded with peak readings in three depressions and near the center of the triangle formed by the depressions. A nearby tree had moderate readings on the side of the tree toward the depressions.
(3) Later in the night a red sun like light was seen through the trees. It moved about and pulsed. At one point it appeared to throw off glowing particles and then broke into five separate white objects and then disappeared. Immediately thereafter, three star like objects were noticed in the sky, two objects to the north and one to the south, all of which were about 10 degrees off the horizon. The objects moved rapidly in

sharp angular movements and displayed red, green, and blue lights. The objects to the north appeared to be elliptical through 8-12 power lens. They then turned to full circles. The objects to the north remained in the sky for an hour or more. The object to the south was visible for two or three hours and beamed down a stream of light from time to time. Numerous individuals, including the undersigned, witnessed the activities in paragraphs 2 and 3.

This document was released to Robert Todd of the *Citizens Against UFO Secrecy,* under provisions of the *Freedom of Information Act.*

Another USAF security officer later came forward with the story of having also witnessed the landed UFO. Dot Street and Brenda Butler co-authors with Jenny Randles of *Sky crash*, a book that deals with this case, revealed his name as Sergeant Larry Warren. Although Warren initially expressing fears for his life, began giving interviews to the media, including BBC TV, Nippon TV, and Cable News Network of Washington, D.C.

Warren adds a few more details. He says that the jeep in which they were riding at the landing site kept failing. He also recounts how a movie camera was pointed toward something which looked like a "transparent round object" hovering just above the ground, about fifty feet in diameter, surrounded by security officers. A bright red light approached from behind the trees, descended silently over the "object" and then exploded in a multiple colored burst of light. Both the "object" and light vanished, leaving in their place a large domed disk with intricate patterns on its surface. Warren and a couple of colleagues approached it, but the next thing he recalls is being back in bed at the Bentwaters Base.

Timothy Good in his fine book *Above Top Secret* cites an astonishing case revealed in 1982 by a former first lieutenant in The Air Force, Dr. Robert Jacobs, now Assistant Professor of Radio-Film TV at the University of Wisconsin. Dr. Jacobs claims that on 15 September 1964, when he was in charge of the filming of missile tests at Vandenburg AFB, California, a UFO was responsible for the destruction of an Atlas missile. "In order to have clear film records of all missile test firings over the Pacific, we had installed a TV camera, affixed to a high powered telescope up on a mountain," Dr. Jacobs reported:

We kept the telescope locked on to the moving missile by radar, and it was while we were tracking one of the Atlas F missiles in this way that we registered the UFO on our film.

Suddenly we saw a UFO swim into the picture. It was very distinct and clear, a round object. It flew right up to our missile and emitted a vivid flash of light. Then it altered course, and hovered briefly over our missile, and then came a second vivid flash of light. Then the UFO flew around the missile twice and set off two more flashes from different angles then it vanished.

A few seconds later, our missile was malfunctioning and tumbling out of control into the Pacific Ocean, hundreds of miles short of its scheduled target.

The film was turned over to two men in plain cloths from Washington. The film hasn't been heard of since. I was told by Major Mansmann , "you are to say nothing about this footage. As far as you are concerned, it never happened!"

It's been 17 years since that incident, and I've told nobody about it until now. I have been afraid of what might happen to me. But the truth is too important for it to be concealed any longer. The UFO's are real. I know they're real. The Air Force knows they're real. And the U.S. government knows they're real. I reckon it's high time that the American public knows it too.

Look magazine, in it's February 1966 issue, tells an incredible story concerning officer Reg Toland of the Exeter, New Hampshire, Police station, who was surprised when a badly frightened young man stumbled into the station at 2:24 A.M. on September 3, 1965. His name was Norman Muscarello and he was almost hysterical. He had been hitchhiking to his home in Exeter when, after crossing the state line into New Hampshire, he suddenly saw a round object 80 or 90 feet in diameter with flickering red lights around the rim, "floating down from the sky" toward him. It "wobbled, yawed, and hovered" overhead, making no sound whatever, and Muscarello, afraid it was going to crush him, dove into a ditch beside the road. But the UFO moved slowly away, pausing to hover for a while over one of the two nearby houses. Then it abruptly flew off.

Muscarello, in a state of panic, hitched a ride to the Exeter police station. Officer Toland wrote down the account, not knowing what to believe. But being a thorough police officer, he called in a patrol car to investigate. The officer who responded to the call, Eugene Bertrand, told Toland that he had just spoken to a woman in a parked car who was terrified because she too had seen a "low flying large round object with flashing red lights."

Len Ortzen, author of *Strange Stories of UFO's*, writing about the same case, said: "Bertrand drove Muscarello back to the place where the latter had seen the UFO. As they walked across the field, Bertrand, who thought the object reported was just a helicopter, saw it for himself. It had returned and was silently hovering about 100 feet above the ground. After a few minutes another patrol car, summoned by radio and driven by David Hunt, arrived. The UFO was still there, and the two officers and Muscarello watched as the brilliant red lights flickered on and off in sequence, casting a scarlet glow over the ground. It finally began to move away, stopping at intervals before it rose and disappeared to the east."

This UFO was seen not only by Muscarello and the two police officers but several other people in the Exeter area also; who reported having observed what was seemingly the same object.

The closing months of 1978 marked some very unusual UFO associated events for the Australian sector of the globe. The tragic case of Frederick Valentich is the starting point of a whole series of visual, radar, and filmed recordings of bizarre flying objects.

"It's approaching from due east of me," radioed the young Australian pilot 50 minutes after he had taken off from Moorabbin Airport, Victoria, on a solo flight in a Cessna 182 aircraft across Bass Strait to King Island. His terse message continued: "It seems to be playing some sort of game. Flying at speed I cannot estimate. It is flying past. It has a long shape ... coming for me right now ... It has a green light and sort of metallic light on the outside."

Valentich was reporting back to Melbourne air flight service controller Steve Robey, after he had radioed a request for confirmation of a large craft with "four bright lights" and had been told that there were no reported aircraft in the area.

"The thing is orbiting on top of me." The Cessna's engine now began to rough idle and cough, and Valentich spoke the last words anyone ever heard from him: "Unknown aircraft now hovering on top of me."

The *Encyclopedia of UFO's*, says, "a loud metallic sound was heard at ground reception for 17 seconds, and then communications went dead. No sign of either Valetich or his plane was ever found, and the mystery remains unsolved to this day."

The La Cronica newspaper in Salta, Argentina, on July 2, 1980, carried a report from Salta about a strange phenomenon seen by no fewer than fifteen people in the Cofico district, near a mountain. After much animated discussion, many residents of Salta had decided that it was all too farfetched to be true when they were jolted by yet another fantastic report. It seems that at 8:15 on July 2, near the Sporting Club in Salta, a boy named Sola suddenly saw the same bright object that the others had seen, above the same mountain. At the same time the boy beheld "at only a few meters distance from him, a strange being about 2 meters in height, hanging suspended in the air, his body emitting a strange luminosity."

This being suddenly spun around like a top, his body remaining otherwise quite motionless. Then he began to rise into the air and finally vanished above the mountain peak! Young Sola stood there dumbfounded, unable to believe his eyes, and then headed for the nearest police station hoping to find somebody there who would believe him.

Meanwhile fresh reports had reached the Salta police from people who said they had seen either a strange craft or strange beings! Many of these were policeman.

Meanwhile, the people of Salta are watching the Andean mountain peaks more closely than ever before, for another glimpse of the strangers from the skies.

Alien Encounters

In a top secret military installation somewhere in the United States, there are those who believe that the government is hiding the remains of an alien spacecraft that mysteriously crashed to earth. With more and more scientific evidence of alien encounters and UFO sightings, the idea of creatures from another planet might not be as far fetched as we once thought. In fact, one of you out there could have the next alien encounter. An encounter similar to the one experienced by police officer PC Alan Godfrey.

Godfrey's experience occurred on November 29, 1980 at about five in the morning, in Todmorden, Yorkshire, and many other witnesses, including police officers, reported a UFO in the vicinity at about the same time. Godfrey described his encounter on worldwide television in these words:

> "I was driving a police car at the time, and in the early hours of the morning I came across what I thought at the time was a bus that had slid across the road sideways. And when I approached the object - I got within about twenty yards of it - and immediately I came across what I now would describe as a UFO.
>
> It was about twenty feet wide and fourteen feet high and was diamond shaped. It had a bank of windows in it and the bottom half was rotating. The police blue beacon was bouncing back off it, as were my headlights. It was hovering off the ground about five feet. And it was very frightening - very frightening.

Jenny Randles, author of The *Pennine,* reports that Godfrey noticed the bushes and trees shaking, which he presumed to be caused by the object. The next minute he found himself 100 yards further down the road and there was no sign of the UFO. Returning to the police station, Godfrey realized that he had experienced a peculiar time lapse.

During several hypnotic sessions later on, he told a bizarre story of having been taken on board a spacecraft! He remembers vague pictures of alien faces. He has flashbacks of being examined by strange alien beings about three feet tall. His life has been changed forever.

Clifford Stone, author of *Let The Evidence Speak,* says: "there is valuable scientific data that proves once and for all that planet earth is being visited by highly-evolved intelligence that is not from this world."

From beyond the boundaries of our perceptions, intelligent beings - both wondrous and terrifying- seem to be abducting people from all over the world, and subjecting them to all kinds of medical examinations.

What is it like to be confronted by a creature whose intelligence and skill is far beyond the comprehension of mankind? Would it be enlightening? Would it be an exercise in terror?

Since the early 1960's, many people report that they have been taken aboard a spacecraft against their will. While the details may vary, most of the stories follow a similar pattern. Most intriguing, perhaps, is the common claim that after the victims observe a UFO, they are surprised to find an hour or two has passed of which they have no recollection. In the days and months that follow, they may experience nightmares, flashbacks, or extreme anxiety. Eventually they began to recall, on their own or through hypnosis that during the missing time aliens abducted them!

Psychiatrists who have examined alleged abduction victims find signs that they have under-gone a severe trauma. Many respected Psychiatrists, such as Dr. John E. Mack, who is a professor at Cambridge Hospital, Harvard Medical School in Boston, theorize, "that if the abduction stories are true, aliens may be conducting long term studies of humans and performing genetic experiments in hopes of creating a human-alien hybrid."

James Cunningham, the author of *The Strong Delusion,* is a Bible expert who takes this one step further. He believes that these aliens may be "fallen angels," who once again, as they did in the time of Noah, are creating a progeny, or cross breed, between "the daughters of man" and "the sons of God." Genesis 6 is a description of just such a happening. (Only this time, their offspring may not be giants, due to the genetic engineering that seems to taking place.)

Cunningham bases his argument on Matthew 24:37 of The New Testament, which states: "As it was in the days of Noah, so it will be at the coming of the Son Of Man."

Cunningham puts forth the theory that the "strong delusion" spoken of in the New Testament, that will fool "even the elect" is fallen angels masquerading as "space brothers."

Judith L. Miller, Ph.D., a clinical psychologist and faculty member of Beaver College believes that "visitors from space are routinely examining human specimens, abducting men, woman, and children, in order to conduct disturbing biological experiments. As in the case of Betty and Barney Hill, who had an encounter of the fourth kind in 1961.This, was sensational front-page news for several years, and the subject of several books. In recent years the story was made into the movie *An Interrupted Journey.*

Here is what happened around midnight on September 19, as Betty and Barney Hill were returning to their home in Portsmouth, New Hampshire, after a vacation in Canada. They were driving south on U.S. Route 3 and had just passed the village of Lancaster, when they saw a moving light in the sky. Near Indian Head it suddenly appeared directly in front of them. Leaving the engine running, Barney got out of the car to observe the object with a pair of binoculars. He could see figures moving behind a row of windows. He said; "the humanoid figures were dressed in shiny black uniforms that looked liked leather and they wore black caps with visors, and moved with odd military precision."

They soon found themselves driving, about two hours later. They drove home, feeling puzzled and uneasy about the two hours of missing time.

They both had flashbacks and recurring dreams over the next few weeks. Finally Dr. Benjamin Simon, a prominent Boston psychiatrist treated the Hills for about six months. During this period, these amazing details were revealed.

A group of humanoid aliens dressed in matching uniforms and military caps, stood in the road, and approached the Hill's car. The leader assured the couple that they would not be harmed. The Hills were then led aboard the disk shaped craft and examined. Samples of hair, fingernails, and scrapings of skin would be taken. Afterward the couple would be returned to their car and allowed to drive home.

One of the most fascinating aspects of the case concerns the "star map" the leader showed Betty when she asked him where he was from. Betty drew this map under posthypnotic suggestion. Several years later an astronomical investigation, based on newly published data not known in 1961, revealed a cluster of stars near two stars called Zeta Reticuli that is amazingly close in configuration to the map drawn by Betty. "This "match" has caused quite a controversy among members of the scientific community." According to *The Encyclopedia of UFO's.*

Over the years hundreds of otherwise credible people have described being somehow immobilized in their cars or homes or wherever, and then taken by UFO occupants into landed UFO's for what appears to be a kind of physical examination conducted while the abductee is stretched out upon a table. What seems to be externally imposed amnesia usually prevents the abductee from recalling the full scenario of his or her experience, which generally lasts an hour or two.

Steven Kilburn, was immobilized after his automobile was pulled off the road as if by some powerful external force. He was then approached by five short, grayish, large-headed humanoid figures. "They're really strange," he recalled under hypnosis. "They're small, below my shoulder. I see the faces and they're white, chalky, and they look like they're made out of rubber." Kilburn's attention was drawn almost hypnotically to his captors' eyes, which he described as "really shiny black. I don't see any pupils or anything, and they're big, they're black and endless. Like they're liquid or something. I keep looking at these eyes looking at me."

Kilburn was taken inside the UFO and placed upon a table where he underwent an intermittently painful physical examination, which included the taking of a sperm sample. Later he was returned to his car and the memory of this traumatic encounter was somehow temporarily blocked.

Budd Hopkins, the author of *Intruders*, claims to "have worked directly with over one hundred people who have apparently had a UFO abduction experience." These individuals, he points out, "Come from every education, social and economic level of our society." He has investigated the cases of "three different abductees who hold Ph.D. degrees." Other abductees he has worked with "include a psychotherapist, a police officer, a lawyer, a diplomat for the United States Government, a farmer, an army officer, a business executive, a well known writer, an artist, a registered nurse, a nice cross section of the community." Hopkins wrote, "I thought that more abduction accounts would continue to come to light, but I was not prepared for the truly vast numbers that apparently exist."

Hopkins points out that "almost everyone who has ever reported a UFO abduction experience has described the behavior of the abductors as peculiarly neutral and objective, displaying neither malice nor human warmth. The general image used by the abductees is that of a laboratory environment, in which they are the tranquilized specimen."

This brings to mind the many television programs, concerning the tracking and monitoring of animals. We have all seen the wild lion brought down with a tranquilizer dart. After which, the animal is examined, tagged for future tracking, and released. Sometimes an electronic device is implanted in the beast, in order to be able to locate the animal at a future time.

Many abductees report an on-going experience, with multiple abductions. In thirteen cases studied by Hopkins, family members from different generations of the same family seem to have been systematically abducted, at varying times and locations, leading one to infer that these abductions represent a genetically focused study of particular bloodlines. Hopkins concludes, "an ongoing genetic study is taking place - and that the human species itself is the subject of a breeding experiment."

Whitley Strieber's x-rays clearly show some type of "device" implanted deep into his brain, and he has "flashbacks" of the long needle like instrument, he says the aliens used to put it there.

Strieber could not explain exactly how he got inside a spacecraft on the night of December 26, 1985. One moment, he said, he was in a wooded area, the captive of two alien intruders; the next, he was whisked from the ground. "I saw branches moving past my face, then a sweep of tree tops. Then a gray floor obscured my vision, slipping below my feet like an iris closing." He soon found himself on a table in "a small operating theater, surrounded by small beings from another world." He has since written several best selling books about his on-going experience. You have no doubt seen him on one of the many talk shows, as he discusses his "relationship" with these "creatures from beyond the realm of our understanding."

In our own galaxy, the Milky Way, there are perhaps 200 billion stars. On some of these it is not unlikely that intelligent beings have evolved and developed civilizations with technologies far superior to our own. Even on this planet, we have a vast difference between technologies. On one side of the world, we have people living in straw and mud huts, while a few thousand miles away buildings rise into the sky. One nation can travel to the moon, while another has never left the ground.

So I would expect that in many of these 200 billion sun systems in our neighborhood alone, we would find everything from the most primitive, to worlds beyond our comprehension.

Calvin Parker and Charles Hickson may have had a glimpse into one of these worlds on October 11, 1973. They were fishing near Pascagoula, Mississippi, when they both saw a bright 20-foot long oval object land nearby. Three occupants emerged and advanced toward them, while the object hovered about three feet off the ground. The creatures seemed to "float" a few inches off the ground as they came near the two terrified men. The sight made Parker faint. Hickson was carried off by one of the beings, while another took the unconscious Parker. They "floated" toward the UFO and entered the craft. Hickson found himself in a very brightly lit room where a large "eye-like" device "examined" him minutely. Afterward Hickson and Parker were both "floated" out of the UFO and deposited back on the riverbank.

Most people who tell of being abducted have described their alien kidnappers as short - about three feet tall - but many emphasize that the leader of the group is slightly taller. Some speak of alien beings with large domed heads set on small bodies, big almond eyes, which do not blink.

The victims all report having felt afraid as the strange looking beings advanced. But they also tell of an almost tranquilizing numbness or paralysis that overtook them, perhaps imposed by the aliens.

While some people report encountering aliens in there homes, many are driving at night on a dark desolate road. They see a lit object flying over or hovering near the car. The car halts, the electrical system is said to fail mysteriously; the radio blares static and then the engine dies. The driver tries frantically to restart the engine, but to no avail. Some victims report staying within the car; others step outside. Despite their terror, they seem unable to run away or call for help.

Four varieties of close encounters of the fourth kind have been distinguished. Type A: Are straightforward encounters where the witness fully remembers what took place. There are no memory blocks, no intervals of missing time, and no obvious reason to doubt that the experience was completely real. Type A events are the best evidence for the reality of aliens, and they are by no means rare.

The second group of contact cases, type B, is quite different. The aliens often make their appearance in the bedroom; the witnesses usually claim to have been wide-awake. Parts of the sequence of events are forgotten. By far the majority of type B cases occur in the home or in the immediate surroundings, and most happen in the early hours of the morning.

The third category of contact reports, type C, involves an experience that is not immediately remembered. The Betty and Barney Hill case is a good example of type C.

In type C contacts, something blocks the witnesses' memory. Occasionally recall is triggered by some normal event, quite often by dreams.

Type C abductions are remarkably consistent. About one in five abductions are type C. They are more subjective than ordinary UFO sightings, but tend to have a higher number of witnesses per case. The most common time for type C incidents to occur is between 10 p.m. and midnight. And most involve young couples driving cars along quite roads.

The last group, type D, is very rare. It consists of those experiences in which the encounter does not seem to involve physical contact. Communication is by means of telepathy, automatic writing, or something of this kind. A psychic encounter.

With all four types, when aliens give a message - the message is almost always one of warnings about the future of the Earth, with hints of nuclear war or impending doom. Occasionally there is some light humor. One abductee suggested to the aliens, that they ought to toast this cross-cultural contact. The aliens did not understand. So he sketched out the chemical structure of alcohol. "How is it that such a highly developed civilization does not use something like this?" the Russian asked. "Maybe if we had used it we would not be so highly developed," came the reply.

Ancient Astronauts

In 66 A.D., Josephus, the Jewish historian noted in his manuscript (The Jewish War) that "before sunset, chariots were seen in the air over the whole country, and armed battalions speeding though the clouds and encircling the city."

Josephus wrote several works of history that rank above any other accounts that have survived over the ages. He was a gifted man of letters, deeply versed in the written source of his own times and earlier days. The *Jewish War* in seven books, cover the period from the conquest of Jerusalem in 170 BC to the tragedy at Mesada in AD 74. Of the strange objects seen in the sky just before Jerusalem was destroyed in AD 66, he writes; "A star that looked like a sword stood over the city and a comet that continued for a whole year. Then again, before the war and the events that led to it, while the people were assembling for the Feast of Unleavened Bread, on the eight of the month Xanthicus, at the ninth hour of the night, so bright a light shone round the temple that it looked like broad daylight; and this lasted for half an hour."

"Then again, not many days after the Feast, on the twenty-first of the month of Artemisium, a supernatural apparition was seen, too amazing to be believed. What I am now to relate would, I imagine, have been dismissed as imaginary, had this not been vouched for by eye witnesses, then followed by subsequent disasters that deserved to be thus signalized. At the Feast called Pentecost, when the priests had entered the inner courts of the Temple by night to perform their usual ministrations, they were stunned with a deafening noise, then was heard a voice as of a multitude crying, "We are departing hence."

One of the first written accounts of a UFO sighting is the following excerpt from an Egyptian papyrus, part of the annals of Thutmose III, who ruled aound 1500 BC:

"In the year 22, of the 3rd month of winter, sixth hour of the day, the scribes of the House Of Life found it was a circle of fire that was coming in the day. It had no head, the breath of its mouth had a foul odor. Its body one rod long and one rod wide. It had no voice. Their hearts became confused through it; then they laid themselves on their bellies. They went to the Pharaoh to report it. His Majesty ordered an examination of all, which is written in the papyrus rolls of the House of Life. His Majesty was meditating upon what happened. Now after some days had passed, these things became more numerous in the skies than ever. They shone more in the sky than the brightness of the sun, and extended to the limits of the four supports of the heaven. Powerful was the position of the fire circles. The army of the Pharaoh looked on with him in their midst. It was after supper. Thereupon, these fire circles ascended higher in the sky towards the south. The Pharaoh caused incense to be brought to make peace on the hearth. And what happened was ordered by the Pharaoh to be written in the annals of the House Of Life. So that it be remembered forever!"

The Roman author Julius Obsequens, believed to have lived in the fourth century AD, drew on Livy as well as other sources of his time to compile his book *Prodigorium Liber,* which describes many peculiar phenomena, some of which could be interpreted as UFO sightings. Here are just a few examples:

"Things like ships were seen in the sky over Italy, at Arpi, east of Rome. A round shield was seen in the sky. At Capua, the sky was all on fire, and one saw figures like ships.

When C. Murius and L. Valerius were consuls, in Tarquinia, there fell in different places, a thing like a flaming torch, and it came suddenly from the sky. Towards sunset, a round object like a globe, or round or circular shield took its path in the sky, from west to east.

In the territory of Spoletium, 65 miles north of Rome, a globe of fire, of golden color, fell to the earth, gyrating. It then seemed to increase in size, rose from the earth, and ascended into the sky, where it obscured the disc of the sun, with its brilliance. It revolved towards the eastern quadrant of the sky"

According to Jacques Vallee, author of *Passport to Magonia*, a term equivalent to our "flying saucer" was actually used by the Japanese approximately 700 years before it came into use in the West. Ancient documents describe an unusual shining object seen the night of October 27, 1180, as a flying "earthenware vessel." After a while the object, which had been heading northeast from a mountain in Kii Province, changed its direction and vanished below the horizon, leaving a luminous trail.

Many unusual celestial events were recorded in Japanese chronicles during the middle Ages. As in Western society, such occurrences were usually considered "portents," often resulting in panics and other social disturbances. For example, On September 12, 1271, the famous priest Nichiren was about to be beheaded at Tatsunokuchi, Kamakura, when there appeared in the sky an object like a full moon, shiny and bright. Needless to say, the officials panicked and the execution was not carried out.

Throughout the 12th century, many UFO's were also seen in England. Here is a classical description from William of Newburgh's Chronicle, of a flying saucer seen in England toward the end of the 12th century:

> *"At Byland, Abbey (the largest Cistercian abbey in England,) in the North Yorkshire Riding, while the abbot and monks were in the refectorium, a flat, round, shinning, silvery object (the word used in the Latin account was "discus") flew over the abbey and caused the utmost terror."*

The European record of possible UFO sightings continued throughout the 14th and 15th centuries. I will list several classic examples:

> *[A.D. 1322] In the first hour of the night of November 4, there was seen in the sky over Uxbridge, England, a pile (pillar) of fire the size of a small boat, pallid and livid in color. It rose from the south, crossed the sky with a sloe and grave motion, and went north. Out of the front of the pile, a fervent red flame burst forth with great beams of light. Its speed increased, and it flew through the air.*

[A.D. 1387] In November and December of this year, a fire in the sky, like a burning and revolving wheel, or round barrel of flame, emitting fire from above, and others in the shape of a long fiery beam, was seen through the winter, in the country of Leicester, England and in Northamptonshire.

[A.D. 1461] On November 1, a fiery thing like an iron rod of good length and as large as one half of the moon was seen in the sky over Arras, France for less than a quarter of an hour. This object was also described as being "shaped like a ship, from which fire was seen flowing." [Jacques Vallee, *UFO's in Space: Anatomy of a Phenomenon, p.9;* Harold T. Wilkins, *Flying Saucers on the Attack,* pp.187, 188]

Monsieur de Rostan, an amateur astronomer and member of the Medicophysical Society of Basel, Switzerland, reported a most unusual sighting. On August 9, 1762, at Luanne, Switzerland, he observed through a telescope a spindle shaped object crossing and eclipsing the sun. Monsieur de Rostan was able to observe this object almost daily for close to a month. He also managed to trace its outline with a camera and sent the picture to the Royal Academy of Sciences in Paris. Unfortunately, according to Harold T. Wilkins, this image - probably the first one ever obtained of a UFO - no longer exists.

The great British astronomer Edmond Halley of comet fame also saw a series of unexplained aerial objects in March of the year 1716. One of them lit up the sky for more than two hours and was so brilliant that Halley could read a printed text by its light. As described by the astronomer, "the glow finally began to wane and then suddenly flared up again as if new fuel had been cast on a fire."

A rare typeset book from 1493, now preserved in a museum at Verdun, France, contains what may be the earliest pictorial representation of a UFO in Europe. Hartmann Schedel, author of the book *Liber Chronicarum,* describes a strange fiery sphere - seen in 1032 - soaring through the sky in a straight course from south to east and then veering toward the setting sun. The illustration accompanying the account shows a cigar shaped form haloed by flames, sailing through a blue sky over a green rolling countryside.

A later chronicler of inexplicable phenomena, one Conrad Wolffhart (a professor of grammar and dialectics who under the pen name of Lycosthenes wrote the compendium *Prodigiorum ac Ostentorum*, published in 1567,) mentions the following events:

> *[A.D. 393] Strange lights were seen in the sky in the days of the Emperor Theodosius. All of a sudden, a bright globe appeared at midnight. It shone brilliantly near the day star (planet Venus) about the circle of the zodiac. This globe shone little less brilliantly than the planet, and little by little, a great number of other glowing orbs drew near the first globe. The spectacle was like a swarm of bees flying round the bee-keeper, and the light of these orbs was as if they were dashing violently against each other. Soon, they blended together into one awful flame, and bodied forth to the eye as a horrible two-edged sword. The strange globe, which was first seen, now appeared like the pommel to a handle, and all the little orbs, fused with the first, shone as brilliantly as the first globe.* [According to Harold Wilkins, This report is similar to modern accounts of UFO formations.]

On the morning of June 30, 1908, something huge and terrifying hurtled out of the sky and exploded over a region called Tunguska in remote Siberia, felling trees as if they were matchsticks and igniting 1,200 square miles of forest. The blast was equivalent to a nuclear bomb, and remains a mystery to this day. Soviet scientists, in a written report, suggest that the blast was from a UFO disintegrating in the atmosphere. The report was based on unusually high radioactivity in the Tunguska soil, and asserts that it came from the spaceship's nuclear engine. Also calculations of the objects trajectory indicate that the occupants of a crashing spaceship deliberately changed course to avoid hitting an inhabited area.

One other event from this era took place before what may have been the largest crowd ever to witness a UFO. In 1917, on the rainy afternoon of October 13, a crowd of some 70,000 people in Fatima, Portugal, watched in amazement as the clouds parted to reveal a huge golden disk spinning and plunging toward the earth. The object gave off vast amounts of heat, drying rain soaked cloths, and water on the ground, in just a few minutes. The disk then climbed back into the sky, and disappeared into the sun. [Time Life Books, *The UFO Phenomenon*, states, "The Catholic Church declared it a miracle."] However, there are striking similarities between this event and many reports of UFO's.

Throughout his long history, man has often wondered whether he is indeed the only intelligent being in his universe and whether life as we know it is confined to the earth alone.

Is there intelligent life in the universe? Are there other living and rational creatures out there besides man? Modern science now seems quite convinced that man is not alone in the universe.

Millions of people have seen objects in the sky that they could not identify, and many thousands have taken the time and trouble to submit written reports about them.

There is no doubt that disc-shaped objects have been seen by a great many honest, sober, and mystified men and woman. So certain are exobiologists of the existence of extraterrestrial intelligence that coded radio messages have been beamed into the vastness of our Milky Way. Moreover, the U.S. space probe Pioneer 10, aiming for a rendezvous 2 million years hence with the star Alderbaran, carried an aluminum plaque bearing a coded message from Earth, including drawings of a stylized man and woman. In addition, for more than two decades, giant radio-astronomy telescopes have been sporadically listening for transmissions from other worlds.

The sighting of strange objects in the sky may actually predate the emergence of modern man. Perhaps the earliest depictions of cylindrical objects resembling spacecraft, with what might be their extraterrestrial occupants, are those carved on a granite mountain in Human Province, China. They have been assigned a tentative age of 47,000 years, predating modern Homo sapiens.

In April 1877, 10,000 residents of Kansas City Missouri watched a "great black airship" hovering in the sky. During this year dozens of reports of strange flying lights and "cigar shaped airships" appeared in U.S. and Canadian newspapers. One headline read: **"Thousands See The Mystic Flying Light."** A California headline on November 22, 1896 read: **"Oaklanders Believe An Airship Hovered Over Them."**

Another wave of "airships" was reported from New Zealand during 1909. For six weeks, from late July to early September, hundreds of people observed cigar shaped airships over North Island and South Island. Sightings occurred in daytime and nighttime.

From the dawn of recorded history, until this present time, strange flying objects have been observed in the skies from very sector of the globe. Yet, the UFO phenomenon remains one of the greatest mysteries of our time.

Moon Mysteries

On the night of July 29, 1953, the late John O'Neil, the distinguished science editor of the New York Herald Tribune settled at his telescope for an evening's observation of the moon.

What he saw stretching across the edge of a great barren expanse known as Mare Crisium was the shadow of a bridge-like structure at least twelve miles long!

Dr. H. P. Wilkins, probably the world's outstanding authority on the moon, calmly announced in August of 1953, that he, too, had clearly seen the same unmistakable bridge-like structure. And the following month, English lunar authority, Patrick Moore, added his voice in support.

In 1912, American astronomer F. B. Harris reported watching a gigantic black object, estimated to be fifty miles in diameter, moving slowly across the moon. He reported that the object cast a shadow on the moon's surface.

On March 30, 1950, Dr. Percy Wilkins, described an oval-shaped, glowing, and seemingly hovering object near the floor of a crater. Dr. James Bartlett, Jr., also spotted the same object.

On February 4, 1966, the Soviet space probe Luna-9 took some startling photographs, which revealed strange towering structures on the moon. Dr. S. Ivanov calculated the structures at fifteen stories high, and spaced at regular intervals, and that they were all identical in measurement.

The Soviet discovery is mind-boggling enough. But, a similar discovery was made by the American space probe Orbiter 2, which took its pictures on November 20, 1966.

Dr. William Blair of the Boeing Institute confirms that they are geometrically positioned. He says the seven spires are not randomly placed.

A soviet space engineer Alexander Abramov concluded shockingly that they form what is known as an "Egyptian Triangle," Says Abramov: "The distribution of these lunar objects is similar to the plan of the Egyptian pyramids. The centers of the spires in this lunar 'abaka' are arranged in precisely the same way as the apices of the three great pyramids."

This startling evidence of intelligence on the moon leads to the reasonable conclusion that this intelligence left its telltale marks on the planet earth!

Apollo 8 astronauts Frank Borman, James Lovell, and William Anders made the first trip to the moon in December 1968. After going into orbit and traveling to the far side of the moon, the Apollo 8 astronauts sighted a huge extraterrestrial object of some kind. They estimated the size of this unusual object - whatever it was - to be about 10 square miles!

On the next orbit they looked all around. But, strangely, the huge object had disappeared. Where could such a vast thing have gone so suddenly?

Speculation ran rampant as to what it could have been: A lunar space station of some kind set up by alien beings. Or was it a huge UFO vehicle of some kind. Or could it have been an extraterrestrial spacecraft merely checking out our Apollo mission to the moon?

V. F. Foster, a British space expert, fully expected that the first men to land on the Moon, would find evidence that extraterrestrial being had been there before us. "In reality", he writes, "such artifact devices may well embody the techniques and principles of superhuman knowledge. Almost certainly, we will soon encounter these objects on the Moon."

Now amazingly it appears that the Soviet Union's Lunokod 2 space probe, a remote controlled robot has discovered just such a startling artifact, while probing the Taurus Mountain region of the Moon!

On February 14, 1973, the Russian robot discovered a smooth stone monolith. The Soviet government reported that it was a sculptured piece of stone, "a plate with a smooth surface." The full UPI news report follows:

> *"The Lunokhod 2 Moon robot parked just over a mile from the Taurus Mountains Wednesday and probed an unusual slab of smooth rock blasted into view by a large meteor, the Tass News Agency said.*
>
> *This one-meter long plate has proved to be a strong monolith. The eight-wheeled robot, which arrived on the Moon*

January 16, was nearly three miles from its landing site on the sea of Serenity.

The plate has a smooth surface, wheras giant stones lying nearby are pockmarked with holes of craters left by tiny meteorites, Tass said."

Many scientists in modern times have come to regard the moon as a cosmic freak of nature. Why? Because by all cosmic laws, the moon should not be orbiting our planet Earth!

Not only is our Moon too large, but scientists also point out it is actually too far out in its orbit to be a natural satellite of our planet. Isaac Asimov states that our Moon is not, as is commonly believed, the true satellite of our planet. He claims that if it were "it would almost certainly be orbiting in the plane of Earth's equator and it isn't." He also rejects the possibility of its having been captured. "But then if the Moon is neither a true satellite of the earth, nor a captured one, then what is it?" (Asimov on Astronomy.)

In his Book, *Space, Time, and Other Things,* Asimov notes: "What makes a total eclipse so remarkable is the sheer astronomical accident that the Moon fits so snugly over the Sun. The Moon is just large enough to cover the sun completely so that a temporary night falls and the stars spring out. And it is just small enough so that during the sun's observation, the corona, especially the brighter parts near the body of the Sun, is completely visible."

The chances of the Moon's being in a position so exact as to equal the disk of the Sun in relation to the planet Earth is so high, as to make it impossible!

The Moon is only 2160 miles in diameter, while the diameter of the Sun is 864,000 miles. The Sun is 93,000,000 miles away, and the Moon is only about a quarter of a million miles away. To quote Asimov, "The chances of the Moon's just happening to be in this position are too incredible to be credible."

All these striking aspects of our Moon's strange orbit around us - its circular, non-equatorial orbit plus the other strange characteristics - seem to make it clear that our nearest neighbor in space is where it is, not by accident.

Dr. Harold Urey summed it up when he wrote before the Apollo journeys: "All explanations now offered are improbable." He holds that to this day. "I do not know the origin of the Moon, I'm not sure of my own or any other's model. I'd lay odds against any of the models proposed being correct."

A well-known American science writer, William Roy Shelton, makes this striking observation in his book *Winning the Moon*, which focuses in on this problem:

> "It is important to remember that something had to put the Moon at or near its present circular pattern around the Earth. Just as an Apollo spacecraft circling the earth every ninety minutes while one hundred miles high has to have a velocity of roughly 18,000 miles per hour to stay in orbit, so something had to give the Moon the precisely required velocity for its weight and altitude. For instance, it could not have been blown out from Earth at some random speed or direction. We found this out when we first began to try to orbit artificial satellites. We discovered that unless the intended satellite reached a certain altitude at a certain speed and on a certain course parallel to the surface of the earth, it would not have the necessary centrifugal force to maintain the delicate balance with the gravity of Earth, which would permit it to remain in the desired orbit". [p.58]

As Shelton further points out, "it is extremely unlikely that any object would just stumble into the right combination of factors required to stay in orbit. Something had to put the Moon at its altitude, on its course, and at its speed. The question is: what was that something?

H. P. Wilkins is the former head of the Lunar Section of the British Astronomical Association, and author of *Our Moon*, one of the most authoritative books ever written on the Moon. This book was published more than twenty years before Soviet scientists postulated their hollow-Moon theory.

Dr. Wilkins (who died at the beginning of the space age) was one of the world's leading lunar experts. He claimed that the Moon could very well be hollow! He wrote: "Everything points to a more or less hollow nature of the moon within some 20 or 30 miles of the surface." He points out that the Moon is only about half as dense as material out of which the planet earth was formed. That is, if one were to take a cubic mile of the Moon and compare its weight with that of a cubic mile of Earth, the latter would weigh nearly twice as much.

In his authoritative study *Our Moon*, the eminent British scientist explains his astounding conclusion:

> *"Long ago it was calculated that if the Moon had contracted on cooling at the same rate as granite, a drop of only 180F. Would create hollows in the interior amounting to no less than 14 millions of cubic miles. However, it is unlikely that the Moon contracted at the same rate as granite; Nevertheless, everything points to the more or less hollow nature of the crust of the Moon, within some 20 or 30 miles of the surface. It thus appears that hidden from us are extensive cavities, underground tunnels, and crevasses, no doubt often connected with the surface by fissures, cracks or blowholes."*

In July, 1970, two Soviet scientists Mikhail Vasin and Alexander Shcherbakov, in a Soviet government publication, claimed that the Moon is hollow and may well indeed be a "spaceship" built by some unknown alien intelligence from a far-flung star system! The report reads in part as follows:

> *"It is quite likely that our Moon is a very ancient "spaceship" with an interior that was filled with fuel for its engines. The hollow interior should also still contain materials and equipment for repair work, navigation instruments, observation devices and all manner of machinery.*
>
> *In other words, the huge "spaceship" carried everything necessary to serve as a kind of Noah's Ark of intelligent creatures on a voyage through the universe for thousands of years.*
>
> *Who were these highly advanced creatures who produced in their mind a spaceship requiring a technology and vision we haven't yet even approached?*
>
> *Abandoning the traditional paths, we have plunged into what may at first sight seem to be fantasy. But the more minutely we go into all the information gathered by man about the Moon, the more we are convinced that there is not a single fact to rule out our supposition. Not only that, but many things so far considered to be lunar enigmas are explainable in the light of this new hypothesis.*

The Soviet scientists insist there is evidence that the hand of alien intelligence in part formed the Moon's inner and outer shells. The outer shell made of rock and dirt and an inner shell, or hull, which they estimate to be 20 miles thick. This inner hull served as space armor to protect the occupants of this huge spaceship in their journeys through the universe.

The Soviet scientists point to the great amounts of metallic elements in the rock samples brought back from the Moon. Vasin and Shcherbakov point out:

> *"This is the perfect kind of material out of which to fashion and reinforce a spacecraft. Such metals were used not only in Spaceship Moon's inner hull but its outside exterior shell to withstand the rigors of their long space odyssey.*
>
> *The metals were chipped from the armor-plate and fused with the loosely packed dirt and rocks in the regions around where they were hit by the super-hot meteorites to form rocks."*

If these Soviet scientists are correct in their surmise that the Moon as an artificially formed spaceship that came here from somewhere else. Driven in and parked in its present orbit, than it may have an internal atmosphere in its hollow regions, probably to sustain life. The Soviet scientists theorize the inner part of this alien world is probably filled with gases required for breathing, and for technological purposes.

Dr. Sean C. Solomon of M.I.T. claims that a study of the gravitational fields of the Moon indicates that it could, indeed, be hollow. Solomon concludes his study, which was published in volume 6 of the technical periodical *The Moon, An International Journal of Lunar Studies:*

"The Lunar Orbiter experiments vastly improved our knowledge of the Moon's gravitational field, indicating the frightening possibility that the Moon might be hollow." [Pp. 147-65.]

Frightening? Yes, because if the Moon is hollow it must have been artificially hollowed out by some alien intelligence - and that would necessarily make it a spaceship!

Buoyed by this knowledge and the evidence that strange moving lights and objects seen on the Moon by astronomical observers, indicated that the Moon was the source of the myriad's of UFOs flooding Earth skies, both the Soviet and the American governments, operating in secrecy, may have launched crash programs to reach this spaceship in Earth skies!

The Headline of the highly respected *Science News Letter* on April 22, 1961 sums up this conclusion: **"The Moon is like a hollow sphere, heavier on the outside than on the inside, according to the data from the Vanguard satellite and new theories about the Moon."**

The first man-made crash directed at the Moon, occurred after Apollo 12 astronauts sent the lunar module ascent stage smashing into the Moon's surface. The shock wave of this hit, staggered NASA scientists. The Moon vibrated for over 55 minutes!

Lunar seismologists anxiously awaited the next big Moon crash. Which came when Apollo 13's third stage crashed into the Moon.

The Moon shook, for 3 hours and 20 minutes! The vibrations of that induced moonquake traveled to a depth of 25 miles. Again NASA seismologists were dumbfounded and could not come up with a satisfactory explanation. However, if the Soviet theory is correct, then such long reverberations are exactly what would be expected. Take any hollow, hard-crusted sphere, especially one that has a metal shell, strike it a hard blow, and it will behave exactly as our Moon behaved.

Our sea-laden world is three-fourths covered with water. The Moon is covered with seas of apparently once molten rock. The question is, where did all the lava come from that created these immense oceans of lava covering a third of the Moon's near side, making it highly reflective?

Still another mystery lurks here, raised by the moon probes - that of the differing far side of the Moon, which has no such lava flows.

These huge circular seas of lava form the familiar "man in the Moon." On our very first trip to the Moon we landed in the left eye - the sea of Tranquility. The sea of Rains makes up the right eye.

Some of these strange dark seas are immense in size. The sea of storms, for instance, is over 2,000,000 square miles!

Most scientists do not believe that the Moon was ever hot inside. The Moon was too tiny a world to generate the kind of heat necessary to produce lava flows. Another strange fact is that they do not appear to be randomly scattered across the Moon's surface; they lay in one quadrant of the Moon's Earth side.

Shcherbakov and Vasin became convinced from this evidence that these lava beds were in fact created by artificial means, by "Moon beings" pouring out huge inner portions of lunar lava and metal not only to hollow out portions of their inner world but to reinforce their metallic armor and prevent it from being damaged by the impact of meteors.

In August 1976 Soviet scientists brought back more samples from the Moon. It was at this time that the Soviet government announced that their scientists had discovered pure iron particles in their lunar samples. These lunar iron particles do not rust!

The Associated Press wire carried this brief but startling Soviet announcement:

> *"Emphasizing the importance of lunar soil samples, an article in Pravda revealed that the first successful automatic mission in 1970 brought back particles of iron that does not rust. Pure iron that does not rust is unknown on Earth. In fact, it cannot yet be even manufactured. Physicists and scientific experts claim they cannot understand how this is at all possible without some kind of manufacturing process being involved. They also point out that it is beyond our present earth technology. Pure iron that does not rust is not found anywhere in a natural state"* [Detroit Free Press, August 24, 1976.)

There is something else discovered about these strange dark regions of the outer Moon that indicates that the hand of alien intelligence was at work here. Walter Sullivan, former science editor of the New York Times, points out: "Furthermore, the sea of Tranquility is covered with material that is considerably more dense than the average density of the Moon, deduced from its gravity. This is the reverse of what one would expect. On earth the lava that flows upwards and out onto the surface is the lighter component - not the heavier fraction." [New York Times, November 9, 1969.]

How do scientists explain how heavier materials make up the Moon's Maria? They cannot. Heavier elements sink, not rise to the surface. The only way it can be explained is as the soviets claim - that the Maria, the dark areas of the Moon, was in some way artificially produced.

Perhaps the biggest mystery of all is the age of the Moon. When our astronauts first went to the moon and brought back rocks for scientists to examine and analyze, scientists never expected to find the Moon rocks far older than the earth, but they did. In fact the rocks were even older than our solar system! Which proves that the Moon came from outside our own solar system, from some other corner of the universe.

The first few trips to our Moon indicated to scientists that the Moon was a dry, dry world. NASA Assistant Director of Lunar Science, at that time, Dr. Richard Allenby concluded: "There is no evidence in the rocks or geochemistry that water exists."

Then came Apollo 15 and the scientific world was rocked back on it's heals again. A cloud of water vapor covering more than 100 square miles was discovered on the surface of the Moon!

Where could a cloud of water vapor 100 square miles have come from? The two Rice University physicists who made the discovery, Dr. John Freeman Jr., and Dr. H. Ken Hills, poured over the data and claimed that the water must have come from deep down inside the Moon!

A few stubborn NASA scientists speculated that this 100 square mile water vapor cloud might have come from the astronauts' urine dump. However, the Rice scientists rejected this (with laughter.)

The impact of the water discovery was momentous. As lunar expert Dr. Farouk El Baz, then a leading NASA geologist, noted: "If water vapor is coming from the Moon's interior this is serious, it means that there is a drastic distinction between the different phases in the lunar interior - that the interior is quite different from what we have seen on the surface." [Science News, October 23, 1971.]

But according to the Soviet hollow Moon theory, our satellite has huge internal areas filled with gases for some kind of atmosphere to sustain life. Could these gases, which might include water vapor, be escaping through cracks and crevasses from the lunar interior out onto the surface of the Moon?

Back in the late thirties, radio engineer Grote Reber was trying out a radical new type of radio telescope experimenting for the Bell Telephone Company in New Jersey. The purpose of the experiments he claimed was to collect and "focus radio waves from space." Riley Crabb of the Borderland Sciences Research Foundation, who relates this incredible account, notes that according to Bell Telephone engineer, Karl Jansky "some thing is happening out there; for Reber turned his radio telescope toward the sky, and did receive signals from the Moon. This was years ago, before such information was classified top secret. [Meeting on the Moon, Borderland Sciences Research Foundation, 1964.]

In 1935 two scientists named Van Der Pol and Stormer detected radio signals on and around the Moon also.

Professor Carl Stormer, a young Norwegian specialist on electromagnetic waves, learned that American scientists had received strangely delayed radio signals from the area of the Moon. Stormer teamed with a Dutch-man, Van der Pol from the Philips Research Institute. They radiated radio call signs of different wavelengths at 30-second intervals. Within three weeks they received these signals back, but with a delay of from 3 to 15 seconds. Scientist all over the world reported similar radio echoes.

Radio signals were picked up and reported in 1956 by Ohio University and other observatories around the world. At the time Ohio University researchers claimed to have received a "code-like" signal from the Moon.

Spaceflight, in its April, 1973 issue, reported that Leonard Carter, the executive secretary of the British Interplanetary Society, claims that a Scottish astronomer by the name of Duncan Lunan discovered a robot satellite placed in orbit around the Moon thousands of years ago!

Professor Lunan discovered this satellite after studying "radio echoes that have been known since the 1920's, and coming from the vicinity of the Moon."

Lunan's research report indicates that computers on this satellite transmit the message whenever they are triggered by radio waves sent from earth at a certain frequency.

This satellite placed in orbit thousands of years ago lay dormant, circling the Moon, until the 1920's, when men on earth began sending radio waves into space. This triggered the device, which then began sending radio waves back to earth.

Much of this information has since been classified, and if the radio signals have ever been deciphered; they have not been made public. One NASA official indicated that one of our manned missions to the Moon captured this strange artificial satellite. Which was then brought back to planet Earth. However, there has been no verification that this occurred.

In 1644 an English Bishop, John Wilkins, wrote the classic book *Discovery Of A New World*, in which he insisted there was a world of living beings inside our Moon! He based this belief on innumerable references in the ancient works in which he was widely read. As a scholar of these works, he had come across references time and time again to the fact that our Moon was a hollow, inhabited world.

Speaking about the Moon, Orpheus, "the Moon man" as he was called, the Greek god (or fallen angel) of wisdom, and son of Apollo (whom the entire Moon landing program was named) tells mankind "that the Moon hath many mountains. And cities and houses built inside the Moon." He also revealed that a day on the Moon is 15 times as long as an Earth day.

Some scholars claim that Apollo, who was the twin brother of Artemis, goddess of the Moon, was in actual fact an ancient astronaut. Some Biblical scholars claim that he was a "fallen angel." Both terms could be correct.

Could it be possible that this "god" and his enigmatic son and all the other "gods" who lived on Mount Olympus were actually alien beings (or fallen angels) from beyond this planet?

Orpheus is responsible for revealing the fact that the world is "egg-shaped," and he is given credit for being the ancient Greeks source of knowledge about "mountains on the moon." Orpheus also revealed to mankind that "man's nature is dual: that he is part of the earth, and part of the heavens."

Orpheus uttered a revealing statement about his fellow "gods" when he said: "These innumerable souls traveled from planet to planet, and in the abyss of space, lament the Heaven they have forgotten."

Since so many Greek "gods" were connected in one way or another with the Moon, could not Orpheus be referring to the very beings that came into orbit around the planet Earth in this Spaceship Moon? And was not Orpheus implying that these "gods" who came from this moon had actually traveled great distances through space?

The Bible states in Genesis 1 that God "placed" the Moon in its present position. Is it possible that angels brought this Noah's ark of a Moon-ship into its present orbit, as part of the creation? Then used this orbiting Moon space station as a base of operations in order to carry out their agenda on the Earth? Were the plants and animals mentioned in this same chapter of Genesis brought to Earth inside this Moon-ship?

Also in this chapter of Genesis, the Bible indicates the Moon would be "the lesser light to rule the night." Could this be why the reflective side of the Moon faces the Earth, in a perfect circular orbit? And could its spacing be for "signs, and for seasons, and for days, and for years," as indicated in the same chapter? Do Angels still live inside the Moon, coming to planet Earth in unidentified flying objects, that the Bible refers to as "The Mystery Clouds?" Do they walk among us now, going about their work unnoticed by most of mankind? Are the many mysterious lights seen on Moon their coming and going?

Maybe this accounts for Plato's strange philosophy of a world of ideas and his belief that thinking is essentially remembering and that the real world of ideas exists not on this Earth, which is merely a "shadow world" but elsewhere. That other "real world" was the world of the "gods" inside the Moon!

Discovery Of *A New World* reveals what the ancient "gods" revealed about the Moon - that it is a hollow world with cities inside of it, and that it is indeed the home of "gods."

We know that one Greek "myth" informs us that Orpheus the Moon man was slain by Zeus, head of the "gods," for "divulging divine secrets."

This is reminiscent of what is written in the Book Of Enoch, when God tells him he is angry because "the sons of god" were having children with the "daughters of man" and that they were "divulging divine secrets," that were unlawful for man to know.

Enoch went on to write that the offspring of these unlawful unions were the giants - the mighty men - of those days. (Which in this author's opinion includes the so-called Greek gods.

Unquestionably the ancient Greeks, even the great thinkers like Aristotle, Plato, Pythagoras, and even Socrates, never really doubted that the "gods" existed - and that they were real beings.

3

Strange Unexplained Phenomenon

Spontaneous Human Combustion

Spontaneous human combustion is a well-documented phenomenon in which a human body ignites and burns without any known contact with an external source of fire. Over a span of some four hundred years more than 300 such cases have been reported.

How could a human body-self ignite, and burn hot enough to be reduced to a pile of ashes? This would require an extreme temperature of more than 3000 degrees! About what you would expect to find at a pressurized crematorium. And what about the unburned clothing associated with some cases? And when you consider that, quite often, furniture and rugs in the same general area were not even singed - the inexplicable becomes the bizarre!

Searching for a pattern in human deaths by mysterious burning, Michael Harrison in his book *Fire From Heaven* considers the possibility that the better educated are spared from seemingly malicious attacks by fire. He cites the case of James Hamilton, who was a professor of mathematics at the University of Nashville, Tennessee. In January 1835, Hamilton was standing outside his house when he felt a stabbing pain in his left leg - "a steady pain like a hornet sting, accompanied by a sensation of heat."

When he looked down, he saw a bright flame, several inches long. "About the size of a dime in diameter, and somewhat flattened at the tip." It was shooting out from his leg. He tried to beat it out with his hands, but to no avail. He did not give way to panic, however. Although what he was seeing was clearly impossible, he accepted that it was taking place. He knew that combustion requires oxygen, so he cupped his hands around the flame, cutting off the supply of oxygen, and the flame went out.

Hamilton's quick action saved him from a fiery attack, and perhaps even death.

Thomas Bartholin recorded one of the earliest well attested cases in 1673. He wrote; "a poor woman was mysteriously consumed by fire in Paris. One evening she went to sleep on a straw pallet and was burned up during the night. In the morning only her head and the ends of her fingers were found; all the rest of her body was reduced to ashes."

The Reverend Joseph Bianchini of Verona in an account dated April 4, 1731, and brought to the notice of the Royal Society of London in 1745 first described the celebrated case of the Countess Di Bandi of Cesena, Italy:

> *"The countess Cornelia Bandi, in the sixty second year of her age, was all day as well as she used to be; but at night was observed, when at supper, dull and heavy. She retired, and went to bed, where she passed three hours and more in familiar discourses with her housekeeper, and in prayers; at last, falling asleep, the door was shut. In the morning, the maid, taking notice that her mistress did not wake at the usual hour, went into the bed chamber, and called her, but not being answered, doubting some ill accident, opened the window, and saw the corpse of her mistress in this deplorable condition.*
>
> *Four feet distance from the bed there was a heap of ashes, two legs untouched, from the foot to the knee, with their stockings on; between them was the lady's head; whose brains, half the back part of the skull and the whole chin, were burnt to ashes; in which were found three fingers blackened. All the rest was ashes, which had this particular quality that they left in the hand, when taken up, greasy and stinking moisture.*
>
> *The air in the room was also observed cumbered with soot floating in it: A small oil lamp on the floor was covered with ashes, but no oil in it. Two candles in candlesticks upon a table stood upright; the cotton wick was left in both, but the tallow was gone and vanished. Somewhat of moisture was about the feet of the candlesticks. The bed received no damage, the blankets and sheets were only raised on one side, as when a person rises up from it, or goes in: The whole furniture, as well as the bed, were spread over with moist and ash colored soot, which had penetrated into the chest-of-drawers, even to foul the linens: Nay the soot was also gone into a neighboring kitchen,*

and hung on the walls, furniture, and utensils. From the pantry a piece of bread covered with that soot, and grown black, was given to several dogs, all of which refused to eat it. In the room above it was moreover taken notice, that from the lower part of the windows trickled down a greasy, loathsome, yellowish liquor; and thereabouts they smelt a stink, without knowing of what; and saw the soot fly around.

It was remarkable, that the floor of the chamber was so thick smeared with a gluish moisture, that it could not be taken off; and the stink spread more and more through the other chambers."

The medical authorities concluded that, 'a naturally caused internal combustion had taken place in her body: Her body was consumed from within, and no external flames were produced that could set fire to the furniture in the room.'

A number of other well publicized cases from that time period included a case reported in the British Medical Journal, April 21, 1888 in which a soldier climbed up into a hayloft in Colchester, England, on Sunday, February 19, 1888, to sleep. He was found completely consumed by fire, while the highly flammable hay around him, both loose, and in bales, was not even scorched.

In a similar case, Wilhelmina Dewar of Whitley Bay, near Blyth, England, was burned to ashes while lying on a bed.
The sheets and bed showed no sign of fire - nor, for that matter, did any part of the house.

The sudden combustion of Mrs. Mary Carpenter, who perished while vacationing on a cabin cruiser off Norfolk, England, on July 29, 1938, took place in full view of her husband and children. She "was engulfed in flames and reduced to a charred corpse" in minutes. No one else was burned and the boat was not damaged.

It was said that water could not extinguish these particular flames; that they burned from within and completely consumed the body. The smoke that issued forth was unlike other smoke, depositing oily, sticky soot on whatever it touched.

On December 15, 1949, the police reported a 53-year-old woman, Mrs. Ellen K. Coutres, had been burnt to ashes in her home in Manchester, New Hampshire. The remains were discovered lying on the floor in a room not scorched by the fire. The fireplace had not been used, and no other source of fire could be found. The associated Press release stated: "There was no other sign of fire, and although the woman must have been a human torch, flames had not ignited the wooden structure of the frame house."

In the 1951 death of Mrs. Reeser of St. Petersburg, Florida, who was found reduced to ashes in a practically undamaged apartment, the St. Petersburg police Chief J.R. Reichert commented as follows: " As far as logical explanations go, this is one of those things that just couldn't have happened, but it did. The case is not closed and may never be to the satisfaction of all concerned."

When found in her apartment, almost nothing remained of the 170 pound woman - only her liver still attached to a piece of backbone, her skull, her ankle and a foot encased in a black satin slipper, and a heap of ashes. These gruesome remains lay inside a blackened circle about four feet in diameter, which also contained a few coiled seat springs. It was an astonishing fact that nothing outside the four-foot circle had been burned.

Spontaneous Combustion is fully accepted as a natural occurrence in the vegetable and mineral kingdoms. Haystacks and stacks of corn have frequently been consumed by the heat generated during the fermentation produced from moisture within them. Spontaneous combustion is fully accepted as the cause of brush fires around the world. Barns, paper mills, stores of explosives, all have gone up in flames from the same causes. In certain circumstances a mass of powered charcoal will heat up as it absorbs air and spontaneously ignite. Bird droppings can produce the same effect.

Travelers who have climbed to the summits of mountains in stormy weather have sometimes observed sparks shooting from their cloths and hair. Their heads can even be surrounded by an electric aureole very much like that of a medieval saint, and is it possible that some people can unconsciously produce a similar affect at ordinary altitudes.

Take the case of Anna Monaro, whose strange glowing hit headlines all over the world in May 1934. An asthma patient in the hospital at Pirano Italy, Mrs. Monaro emitted a blue glow from her breasts as she slept. This emanation lasted for several seconds at a time, occurred several times a night, and continued for a period of weeks. Many doctors and psychiatrists observed the phenomenon. None could explain it.

Mrs. Olga Worth Stephens, age 75, of Dallas, Texas, was sitting in a parked car in October 1964 when witnesses saw her burst into flames. She was reduced to ashes before anyone could come to her rescue. Yet, the automobile was not damaged and Fireman said, "that the car contained nothing that could have started the fire!"

In one of the most spectacular cases of spontaneous combustion, Sven Pehrson, 24, while lying sunbathing on the beach, in 90-degree weather, suddenly burst into flames - incinerating him in a matter of minutes!

Ski instructor Sven and his new bride Pia were on their honeymoon. Swimming in the ocean and basking in the sun. Pia told police "that in the afternoon we were just lying on our blankets, laughing and holding hands. All at once I heard a strange hissing noise and Sven got a wild look in his eyes and I knew that something was terribly wrong.

"Then there was a loud *whoof* and he was totally engulfed by fire! One second he was smiling and talking to me and the next second he was gone."

Stunned sunbather Antonio Iziar witnessed the deadly spectacle from his blanket just a few feet away.

"I was watching the young couple because they were so happy and she was so lovely," Antonio, 56 told reporters. "Suddenly the boy just caught on fire, and in a few seconds he was a ball of fire!"

Astonished scientists hope the incredible case will help them determine why certain people suddenly burst into flames for no apparent reason. Scientists remain mystified by the chilling phenomenon of spontaneous human combustion in which dozens of people each year explode into flames internally for no apparent reason.

Victims simply self ignite and are reduced to ashes, sometimes leaving behind an extremity, usually an arm or a foot, and always, a lot of unanswered questions.

Mysterious Vanishings

There is no stranger case in the annals of the Royal Canadian Mounted Police, than the village that vanished! The entire village, men woman, and children had inexplicably disappeared!

Strange Unexplained Phenomena

About 500 miles Northwest of the Mountie base at Churchill, at Lake Angikuni, a village of Eskimos had lived for many years. This was a great spot for them to survive.

Here was good hunting and fishing, and far enough from civilization that they could live as they pleased. Trappers, such as Joe LaBelle, knew the Eskimos well, and would often stop to visit and swap furs, eat Caribou, and rest.

On a cold November day in 1930, Joe approached the village, and shouted a greeting.

No reply of any kind - no barking dogs - no children running to greet him - only silence.

Joe spent about an hour in the deserted village. He examined all the tents and huts, and searched for any clue to the strange disappearance of the complete village.

Pots of cold food, which had been cold for some time, hung over cold cooking fire pits.

In one hut was a sealskin garment, the ivory needle still sticking in the garment where someone had abruptly ceased the mending. Three Kayaks, neglected for some time, lay battered by wave action on the beach.

Labelle went to the Mounted Police for help. He led them back to the village, where they began a search for the missing tribe.

The police found the Eskimos' prized rifles standing in their normal place. No Eskimo would start on an extended trip without his rifle, yet here were the guns - and the Eskimos were gone!

Near the deserted village, the police found seven dogs. All had been dead for some time.

Dead from starvation, according to the Canadian Government Pathologists.

No clue was ever found as to where the Eskimos had gone. They just simply vanished, leaving behind their clothing, food, guns, and dogs!

Months of investigation among the other tribes in the area brought no trace of any member of the missing village. And no member of this village has ever been seen or heard from since!

And in the files of The Mounted Police it remains an unsolved mystery.

Matthias Grimsson leaped out of an airplane with four other skydivers—and mysteriously vanished into thin air!

Aviation officials have been called in to investigate the strange disappearance. Search and rescue teams were organized by the authorities to search every inch of land in the area. However his friends who were with him during his last jump, all say he was lost to sight before he hit the ground!

The story reported by the Reykjavik news in Akranes, Iceland, is one of the most bizarre news items ever reported.

"He faded, He vaporized," a badly shaken companion jumper told authorities. "He jumped and fell about 300 feet, then he just seemed to dissolve in mid-air!"

"There wasn't much cloud cover and the sun wasn't in my eyes, I know what I saw. They can search all they want for Matthias but they aren't going to find a trace."

"Aviation officials are reluctant to accept eyewitness accounts of the tragedy because there is no logical explanation for the strange disappearance. But the man's family say they believe Grimsson's friends, and are pressuring the government to release more information about recent experiments involving U.S. and British aircraft in the area." The Reykjavik news reported:

"They've been conducting tests, that much we know," said Grimsson's brother, Gunnar. "We want to know what sorts of new weaponry they've been testing. Maybe they can explain how a man can be a living, breathing being one second and nothing but a cloud of vapor the next." The Grimsson family has been enthusiastic skydivers for seven years before he disappeared and he believes there is little chance he met with an emergency that he could not handle.

Grimsson's dad said, "he was in sight of four people at the time he vanished.

"They didn't see him fall, he never passed them, and there are no remains on the ground. This man disappeared in some other mysterious fashion and we want to know how it happened."

However the case remains unsolved, and no body has ever been found, and no one has seen or heard from him since that fateful day in mid July.

As strange as this story is, this is not the first man to vanish into thin air. Our next account is even more bizarre, and stranger than you can imagine!

Blimp L-8 had taken off from Moffett field about 6 a.m. this calm early morning. Lieutenant Earnest D. Cody and Ensign Charles E. Adams at the controls. At around 8 a.m., Cody sent a radio message saying that he had seen a suspicious oil slick. "I'm taking the ship down to 300 feet for a closer look," he said. "Stand by."

The L-8's position at that time was a few miles off Farallon Island. Two armed patrol boats had been alerted and were observing the blimp. Two fishing trawlers in the area also watched the airship circling overhead.

At 8:05 a.m. the control tower at Moffett Field tried to contact the blimp, but had no response.

At 10:40 a.m. the blimp had floated along the beach where two fishermen tried to catch it, and saw that it was empty. About 30 minutes later, the airship settled to earth in a street in Daly City, close to San Francisco.

When the salvage crew investigated the Blimp's gondola, they found all equipment in position. The rubber raft and the Parachutes were in their place. The only missing items were two bright yellow life jackets, which the crew was required to wear. The door was shut and latched from the inside and nothing had been damaged. There was no sign that the gondola had hit the water. There had been no fall or splash observed by the fishermen who were watching the Blimp. No bodies or life jackets were ever seen or recovered after an extensive search. Even though the jackets were bright yellow.

To this day no one has any idea what happened to Cody and Adams, and their fate remains an unsolved mystery of two men who vanished into thin air.

Mystery also surrounds the last flight of George Guynemer who by September 1917 was the leading French ace with 54 German planes to his credit. The French called him "George the Miraculous." Guynemer, took to the air September 11, with two other members of his celebrated Stork Squadron. Catching sight of a German two-seater in the distance, he signaled to his companions to stay behind as cover and lookout. He banked his plane into a cloud, and was never seen again! A search up and down the lines and in the sky above the clouds revealed nothing. Guynemer never turned up again.

Had he been shot down, the Germans would have claimed his death, for he was a real prize. But no announcement was ever forth coming.

In a similar case Benjamin Bathurst, British envoy to Vienna during the Napoleonic wars, in 1809, was on his way from Vienna Via Berlin to Hamburg. Bathurst stopped at the small town of Perelberg to change horses on the coach. Bathurst took a step from the coach and vanished just as completely as if the earth had opened up and swallowed him. There were three eyewitnesses to this strange event: Bathurst's Swiss servant, along with two fellow travelers.

A German chronicler of the events writes, "The disappearance of the English ambassador seems like magic; for he disappeared in plain view of three other people leaving no trace behind. And he was never seen or heard from again."

An interesting case of someone who disappeared and reappeared several times before vanishing a final time was that of Private Jerry Irwin of the United states Army. The first time he vanished in plain view of two fellow soldiers, who reported it to the officer in charge. A search was made of the entire camp, but Irwin could not be found. Then all of a sudden he reappeared at the camp. But as he was being questioned about his whereabouts - he disappeared again!

A second search was made, but he was not to be found. A few hours later he suddenly reappeared once more. Neither time was the soldier able to say where he had been. Then on August 1, 1959, Private Jerry Irwin disappeared for good.

Two other fellow soldiers, Bill Monroe, and Charles Kilp, said they saw him walking across the lawn toward the officers quarters, when in "mid-stride" he vanished forever!

A few miles from Gallatin, Tennessee, on September 23, 1880, David Lang vanished from the face of the earth, in full view of his family and friends!

On that particular afternoon, Lang's two children, George, eight, and his daughter, eleven year old Sarah, were playing in the yard as their mother and father came out of the house.

Mr. Lang could see a horse and buggy coming down the long drive, and knew that it was his old friend Judge August Peck, and his brother-in-law, coming to pay a visit. He waved at the Judge and took six or seven steps when he disappeared in full view of all those present.

Mrs. Lang screamed, Judge Peck and his brother-in-law, jumped from the buggy, and ran to the spot were Lang had been standing only moments before. There was not a bush, not a tree, nor was there any hole that could be found. They searched for about an hour, and nothing could be found. Mrs. Lang was led to the house screaming. Judge Peck started ringing a big bell that stood in the side yard to alert neighbors that something was wrong. By nightfall several dozen people were on the scene. They lit lanterns and torches and searched every inch of that yard in which David Lang had been seen a few hours before. They used rods and poles to poke the ground - looking for a hole or a cave in, but the ground was firm and had not been disturbed.

David Lang was gone! He had vanished in full view of his wife, his two children, and the two men who came to visit. One second he was there, walking across his yard in bright sunlight, the next instant he was gone! Every one agreed that he did not fall or slide way - he just vanished in the twinkling of an eye.

Subsequent examination of the yard found it to be supported by a thick layer of stone. Without sinkholes or caves. His wife, who lived for many years after the event, never held a memorial service. She always felt that he might somehow return. But he never did.

About seven months after Lang had vanished, the children noticed that at the spot where he had last been seen, there was a circle of stunted yellow grass some fifteen feet in diameter. The children claimed that at times they could hear the faint voice of their father calling for help, in the area of this circle. However, nobody else could hear anything, and it was felt that the children were suffering from the loss of their father, and might be hearing things that were not there.

This case is very similar to that of Charles Ashmore's disappearance. This 16 year old boy set off from the family farmhouse one snowy winter evening, apparently to fetch water from the nearby spring. When he failed to return his father and sister went to search for him. They found that his footsteps in the snow stopped short about halfway along the path to the spring. Young Ashmore never reappeared, though his voice was heard in the immediate area for several months after.

Just as perplexing was the 1889 case of eleven-year old Oliver Larch of South Bend, Indiana, who vanished on his way to a well on Christmas Eve.

Oliver Larch lived with his parents on a farm, a few miles out of the city. An old-fashioned Christmas party was under way.

The family minister and his wife were there, along with a circuit judge from Sound Bend and an attorney from Chicago who had been a friend of the family. It was a delightful get-together, which had become an annual event for this little group.

Outside, the snow had stopped falling. It was about five inches deep; A soft fluffy blanket that lay as it fell, for there was no breeze on this black, star-less night.

A few minutes before eleven o'clock, Oliver's father noticed that the gray bucket that held the drinking water needed filling. He asked

Oliver to run out to the well and bring in a bucket of fresh water. Oliver slipped on a pair of overshoes and went out the side door as his father went back into the parlor to be with the guests.

A few moments after the boy set out, the assembled guests heard Oliver yell for help. They ran out the same door Oliver had used. Mr. Larch brought a kerosene lamp, which lit the area. But they found only his footprints in the newly fallen snow - the trail began at the door and abruptly ended only halfway to the well.

The group could hear Oliver screaming for help. The witnesses afterwards agreed that the cries for help were coming from overhead. Somewhere up in the sky - Oliver Larch was in mortal fear - and his screams were getting fainter and fainter.

The bucket lay on its side in the snow at the place where the footprints ended. There were no other marks of any kind in the snow.

Oliver Larch never reached the well. He never turned aside in the snow. His footprints extended a little more than thirty feet from the side porch before they came to an abrupt end.

The boy weighed about seventy-five pounds, far too much for an eagle to lift.

The mystery of Oliver Larch is as baffling today as it was on that Christmas Eve.

In another strange case of mysterious disappearance, Owen Parfitt vanished from the chair in which he was sitting.

Parfitt, age 70, had been bedridden for a number of years. He could not move his body without assistance, nor could he walk.

His sister Susannah Snook, had lifted him from his bed, and set him in a rocking chair next to the door. He was wrapped in a blanket, and in a good mood.

His sister was away making the old man's bed for no more than ten minutes, and maybe a little less.

When she returned he was gone from his chair. The blanket was found on the floor, next to the chair. But Owin Parfitt was never found or heard from again.

While driving from Gila Bend To Ajo in 1958, Daniel Kramer of Phoenix, Arizona, blew a tire. He had no jack, no water, and no air-conditioning.

It was one of those hot afternoons in the Arizona desert. The Sun caused shimmering heat waves to dance just above the scorching asphalt, for as far as one could see, down the lonely deserted road.

Kramer was weak from the heat, very thirsty, and hadn't seen another vehicle all day, and there wasn't a sign of life in any direction!

All of a sudden - Kramer heard someone ask if he needed any help.

"I was dumb-struck," said Kramer. "When I turned around, a tall, slender man was standing beside me and his old Ford coupe was parked on the gravel shoulder behind my car."

Because of the silence Kramer would have certainly heard a car drive up.

"But I heard nothing before he appeared," the astonished Kramer recounted. "No humming of tires, no grinding of gravel, no sound of a car door opening and closing. Yet there he stood."

The stranger got a jack and lug wrench from his car, and changed the tire for this bewildered traveler.

"I tried to talk to him, while he worked," Kramer said later, "but he didn't answer."

"When the man finished, I offered to pay him." Kramer said, as he shook his head in disbelief. "But he told me to just help someone else someday when they're in need."

After the stranger had said those few words, he got into his car and drove out onto the highway.

But as Kramer watched the mysterious car "It simply disappeared!"

One moment it was there, and the next split second, while he had his eyes glued to the back of the strange car, it was gone! Who was this stranger that suddenly appeared, and then just as suddenly disappeared? We may never know.

The U.S. Navy has been credited with many feats, but one of the strangest is that it briefly vanished one of its own warships during World War II.

The Philadelphia Experiment, as the alleged disappearance is popularly known, is said to have occurred at that city's port in October 1943. According to one person who claimed to have observed the incident, an American destroyer at dock was fitted with large magnetic field generators.

When the generators were turned on a "green haze" extending a hundred yards from each side of the ship suddenly surrounded the ship. Then the vessel disappeared, leaving only the impression of its hull against the ocean.

A few minutes later, after the generators were shut down, the ship reportedly appeared at the naval base in Norfolk, Virginia, more than 200 miles away. Then it vanished again and reappeared in Philadelphia.

Strange things began to happen to crewman who had been aboard the USS Eldridge, the fully manned destroyer escort that was involved in this secret wartime Navy experiment.

According to Carl Allen, an eyewitness, two sailors burst mysteriously into flames. But the strangest aspect of the experiment was that many of the crewman flickered back and forth into invisibility.

Dr. Samual Kohl, many years later, said he had treated several of the affected men. "They would fade in and out of invisibility while they were being treated - right in plain view of everyone in the room."

Dr. Kohl said he had been sworn to secrecy, and had kept the whole thing to himself, until he was near death in 1972, when he decided to tell what he knew about the bizarre experiments.

Most of the sailors who continued to "fade in and out," went insane." Said the doctor. Several other witnesses since also confirmed this fact.

Information about the Philadelphia Experiment began to come to light in 1956 in a series of letters sent to Maurice K. Jessup, an astronomer and author of a book called *The Case For The UFO.* An eyewitness sent the letters.

Jessup was about to make the whole story public, and had just called a member of the press to meet him for a briefing. Jessup did not show up for that appointment, and was found dead in his car the same day. A hose was connected from the exhaust pipe into the car. Jessup was lying dead on the seat. His papers, letters and brief case were missing, and have never been found.

The Navy's office of Naval Research has steadfastly denied any such experiment ever took place. However, a number of books have been written, and several movies made about the so-called Philadelphia Experiment.

The pages from the logbook of the USS Eldridge for the time period of the alleged experiment are strangely missing, and requests for information on this ship, for that time period, go unanswered.

What of the steamboat that sailed around the bend to oblivion in broad daylight? The Iron Mountain shoved off from Vicksburg on a bright June morning in 1872. Moments later she had gone around a bend in the river and vanished from sight. The Iron Mountain, 180 feet long and 35 feet wide, with 55 passengers and crew, was never seen again. No wreckage was ever found, and there were no survivors. The big steamboat just vanished without a trace.

On December 14, 1928, the Danish ship Kobenhaven suffered the same inexplicable fate. The ship steamed slowly out of the harbor, past several other vessels - and vanished. Incoming ships, which should have sighted her, saw neither the vessel nor any trace of her.

In July of 1909, the Waratah steamed out of Durban, South Africa, with a full cargo, and two hundred and eleven people aboard. She disappeared as completely as if she had sailed off the face of the earth. A sixteen thousand ton steamer equipped with lifeboats, flares and rockets, and in good weather, vanished without a trace in a shipping lane where eight other vessels would have seen her rockets. The search lasted for weeks, but not a plank, not a life preserver, nor a body was ever found.

The U.S.S. Cyclops left Barbados on March 4, 1918, bound for Hampton Roads, Virginia, with a crew of eighty-eight, and a full cargo. Like the Waratah, she also vanished without a trace in an area that was well traveled. This ship had full radio equipment, and a number of other ships would have heard a distress call. She sent no messages. No trace was ever found. She just sailed over the horizon, and was gone forever.

The list of aeronauts who have vanished with their craft is long. But how can we explain the disappearance of fliers and planes on short flights over heavily populated terrain?

Albert Jewel was simply going to take a little flight in his airplane over some of the most heavily populated areas in the nation, on a clear day in October 1913. He has not been heard from since. Search parties, which included the military, found no sign of wreckage. The case remains unsolved.

Captain Mansell James was about to land at Mitchell field, Long Island, but like Mr. Jewel, something detained him. He was last seen circling the airport - but then he was gone. Captain James and his plane had flown away to nowhere, and in the ensuing years they have been joined by many others.

There was the unexplained disappearance of the French Latecoere, a huge flying boat capable of landing on the ocean, which vanished over the North Atlantic on August 1, 1948 carrying fifty-two people to oblivion. No radio messages, no wreckage, and no life jackets.

The Star Tiger with twenty-three passengers and six crewmembers were just a short distance from their destination in

Bermuda. It had radioed, "On course, good weather, and no trouble." But then it just simply vanished in the night. Ten ships of the U.S. Navy, joined with thirty British and American planes in the fruitless search. The disappearance was so sudden that the crew had no time to radio for help, and so complete that no trace was ever found.

It is noteworthy that many missing ships and planes disappear in a comparatively small area of the Atlantic north of Bermuda and bordered on the west from Florida to Virginia. This area is called the Bermuda triangle. Perhaps the strangest of these occurred on the afternoon of December 5, 1945.

Five TBM Avenger propeller-driven torpedo bombers left the base on a routine flight. They took off from the Naval Air Station in Fort Lauderdale, Florida. They were to fly 160 miles east over the ocean, forty miles north and then southwest back to the base.

One plane carried two men; the others carried three each. All were equipped with the best radio and navigational equipment, self-inflating rafts, and each man wore a life jacket.

The first sign of trouble came at 3:45 in the afternoon. The radio message said: "Can't be sure where we are. Can't see land. I'm not sure of our position." All five navigators seemed to be lost at the same time. At four o'clock the tower heard the panicky flight commander turn the command over to another pilot. The last message came in at four twenty-five: "Still not certain where we are - the sea looks funny." The voice trailed off into silence, which has engulfed both men and planes ever since.

A big Martin Mariner flying boat (which could land on the water) with a crew of thirteen, and loaded with rescue and survival equipment, took off to find the lost planes, and guide them home. The lost aircraft should have been within 10 minutes of the base.

About ten minutes later the base radioed the big flying boat to check its position. No reply. The Mariner had also disappeared in the same general area.

240 planes were involved in the search, which covered the whole area from Florida to the Bahamas. Not one of them ever found a trace of the now six missing planes or their crews.

Before the search was finally abandoned it had developed into the greatest air-sea rescue attempt on record, with 21 ships, 300 planes, and twelve land parties, that searched the shores of the mainland, the seas, and the islands for weeks, seeking some clue to the missing planes and crews.

The Naval Board of Inquiry said: "We are not able to make even a good guess as to what happened."

History also records several cases of vanishing armies, including four thousand trained and fully equipped troops who camped one night beside a small stream in the Pyrenees during the Spanish War of Succession. Next morning they had a rendezvous with destiny, for they broke camp, formed ranks and marched into the foothills to disappear from sight, from that day to this. No traces of the army, its equipment, or any of the men were ever found.

It was a fate shared with the 650 French colonial troops who were marching toward Saigon, Indo-China, in 1858. Trudging along the highway across open country only fifteen miles from the city, they vanished as completely as if they had walked off the earth.

A third army commanded by Colonel Li Fu Sien of China, had just made camp in an area of the rolling country south of Nanking. Two hours later his aide reported that the troops did not respond to radio calls. Their guns were found stacked beside their little cooking fires, but the troops were gone. They could not have deserted in mass across the open country without being detected. There was no battle. And not one of the 2,988 missing men has ever been seen or heard from since.

One of the strangest vanishing cases in history involved a twin-engine marine plane with thirty-two persons aboard which crashed on Tahoma Glacier, in 1947. Two days later searchers reached the plane - but there were no bodies! It was evident that no one could have survived the crash; yet, there was no sign that anyone had been aboard the plane when it crashed! What had happened to the 32 men who had been aboard the plane? A reward of five thousand dollars was offered for discovery of any of the bodies - or body parts - but not one of them has ever been found.

Psychic Dreams

The phenomenon of precognitive dreams is the most common type of all inexplicable mental experiences. Because of its very nature, it is the least understood and one of the most difficult to study. The wonderland of the human mind contains many mysteries, and science has unlocked only a few of the many secrets that lurk there.

On January 29, 1963, Mrs. John Walik of Long Beach was having a nightmare at about three in the morning.

As she was awakened with fright, she realized that the aircraft she saw in her dream was the same one in which her husband was a crewmember.

In the dream, the airplane leveled off for a landing, hooked a wing into the earth and cart wheeled - in a ball of flames!

She phoned the Slick Airways that morning, but the office could only tell her that no plane crashes had been reported.

The big four-engine constellation that she had seen in her dream was not due back for a few days.

Mrs. Walik could not rid herself of that dream, and she felt it must be a prelude to disaster. She told her family, neighbors, and friends about the dream and her fears.

On Sunday morning, February 3, 1963, she again called the office of the airline for which John worked. They again told her that no trouble had been reported, and in fact, the plane was due in at any minute.

Mrs. Walik was still on the line when her husband's plane, coming in for a landing, caught a wing on the ground, and cart wheeled in a ball of fire. Just as she had seen in her dream two nights before!

The headline on February 4, 1963 of the Long Beach Independent Press read: **"Mate's Plane Crash Seen In wife's Dream."**

Contrary to popular belief, dreams of this nature are not uncommon. History is filled with dreams of foreboding that eventually become reality. The English journalist William T. Stead, editor of the Pall Mall Gazette, recorded the following dream of a brother's murder in 1840. Persons intimately acquainted with George Northy, the brother of Hart, the murdered man, had given him the story.

George and Hart had never been separated since their birth until George became a sailor. Hart at the same time joined his father in business. In February 1840, George Northey's ship was lying in port at St. Helena. While he was there George had a strange dream, which he himself related as follows:

> "I dreamed that my brother Hart was at Trebodwina Market, and that I was with him, quite close by his side, during the whole of the market transactions. Although I could see and hear everything which passed around me, I felt sure that it was not my bodily presence which thus accompanied him, but my

shadow, or rather my spiritual self, for he seemed quite unconscious that I was near him.

I felt that my being thus present in this strange way betokened some hidden danger which he was destined to meet, and which I knew my presence could not avert, for I could not speak to warn him of his peril. My terror gradually increased as Hart approached the hamlet of Polkerrow, until I was in a perfect frenzy, yet unable to warn my brother in some way and prevent him going further...

I suddenly became aware of two dark shadows thrown across the road. Two men appeared whom I instantly recognized as notorious poachers, who lived in a lonely wood near St. Eglos. The men wished him 'good evening,' civilly enough. He replied and entered into a conversation with them about some work he had promised them.

After a few minutes they asked him for some money. The elder of the two brothers, who was standing near the horse's head said:

' Mr. Northey, we know you have just come from Trebodwina Market with plenty of money in your pockets, we are desperate men, and you are not going to leave this place until we've got the money; so hand it over.'

My brother made no reply, except to slash at him with the whip and spur the horse at him.

The younger of the ruffians instantly drew a pistol and fired. Hart dropped lifeless from the saddle, and one of the villains held him by the throat with a grip of iron for some minutes, as though to make assurance doubly sure, and crush out any particle of life my poor brother might have left.

The murderers secured the horse to a tree in the orchard, and having rifled the corpse, they dragged it up the stream, concealing it under the overhanging banks of the watercourse. They then carefully covered over all signs of blood on the road, and hid the pistol in the thatch of a disused hut close to the roadside;
then, setting the horse free to gallop home alone, they decamped across the country to their own cottage."

George Northey eventually reached port, only to find that his brother Hart, had been murdered! Exactly as he had seen in the dream!

George Northey arrived from St. Helena and declared that the pistol was hidden in the thatch of the old cottage close by the place where his brother was killed, and that the Hightwood brothers had hidden it there.

The Hightwood brothers were arrested; a search was made of their cottage, which revealed bloodstained garments. The gun was found exactly in the place George said it was.

"How did you know?" he was asked, after the gun had been found. "I saw the foul deed committed in a dream," he declared.

Many such dreams have been recorded through out history. Many more have gone unrecorded. One of the most well knows dreams of disaster were that of Mark Twain's in the late 1850's.

Mark Twain had been sleeping at his sister's house in St. Louis, when he had a very vivid dream. He dreamed he saw his brother's corpse lying in a metal coffin, with a bouquet with a single red rose in the center, placed on Henry's chest.

A few weeks later, his brother Henry, was killed in a boating accident.

When Mark Twain came to say his last farewells to his brother, he found the body lying in a metal coffin, just as he had seen in the dream. However, the bouquet was missing. But as Twain stood beside the body a woman came into the room, and placed a bouquet on Henry's chest. They were white flowers, with a single red rose at its center!

The dream itself is a texture woven of space and time inside which we find ourselves. During the dream we believe we are awake, in the same way that we believe we are awake when we truly are.

Let us consider the strange case of Mrs. Winnie Wilkinson of Sheffield, England, who lay down to take a nap one afternoon in the summer of 1962. Mrs. Wilkinson said that she dreamed of hearing a heavy and persistent pounding on her front door. When she opened the door in her dream, an excited woman greeted her whom she had never seen before. The woman told Mrs. Wilkinson that Mr. Wilkinson had just fallen from a scaffold and was badly injured, and he wanted his wife to come at once.

At that point Mrs. Wilkinson awakened, and was quite upset by the strange dream. She noted the time of 3:12 p.m., and phoned his employer and was assured that there had been no accident.

On the following day exactly at 3:12 p.m., Gorden Wilkinson was killed when the scaffold on which he was working suddenly collapsed.

Ever since humans discovered that they dream, they have wanted to know why, how and what it meant in relation to their waking existence.

Before we can discuss dreams we must fully understand the nature of their companion, sleep, without which dreaming would have no objective reality. Sleep is by no means a total disconnection from activities as is shutting of a machine. During sleep the conscious mind rests temporarily, while the unconscious part of the mind continues to function. Vital activities of the body are carried on automatically.

One of the best-known authorities on dreams is Calvin S. Hall, director of the Institute of Dream Research at Santa Cruz, California, and the author of a number of books on dreams. "A dream is a succession of images, and sounds, which are experienced during sleep," Dr. Hall writes in *The Meaning of Dreams*. He views dreams very much like stage players. "A dream commonly has one or more scenes, several characters in addition to the dreamer, and a sequence of action and interactions usually involving the dreamer. It resembles a motion picture or dramatic production in which the dreamer is both a participant and observer. Although a dream is an hallucination, since the events of a dream do not actually take place, the dreamer experiences it as though he were seeing something real." Erich Fromm, an internationally recognized authority on dreams, considers the dream state, a state of heightened receptiveness, necessary to convey to humans "certain information which they would normally not accept because of the nature of their psyche."

The unconscious part of the mind is allowed free range of expression, while the physical body continues to function on a reduced scale." A degree of filtering is involved in order to make the material acceptable to the dreamer. The psyche, or spirit, is now free to send symbolic material upward toward the conscious level so it can be understood when the dreamer awakes.

We thus have a two-way traffic, external material being received and sorted out, and internal material being sent out to the dreamer's conscious, in order to call his or her attention to certain conditions of which he or she is not normally aware.

In the spring of 1915 the eminent British lecturer, I. B. S. Holbourne, was returning from a highly successful lecture tour of the United States. He had booked passage on the big Cunard liner *Lusitania*.

Sometime during the night of May 7, 1915, the professor's wife, Marion, dreamed that the ship was sinking. In the dream Mrs.

Holbourne saw her husband get into one of the lifeboats, which was safely launched before the big liner slid beneath the waves.

At breakfast that morning she discussed the dream with her family and friends.

Hours later the news came that the mighty *Lusitania* had been sunk by action of a German submarine off the Irish coast with heavy loss of life. Professor Holbourne, they learned later, had helped many others into life jackets and boats before he was finally ordered into a lifeboat himself. The Professor was a survivor.

Dreams due to physical discomfort or environmental pressures are not as vivid as psychic dreams and are more easily forgotten. Not so with psychic dreams. One is rarely able to shake them. Some psychic dreams are so strong that they awaken the dreamer, and the dream remains clearly etched into the memory for long periods of time after the dream.

Such was the dream of a father who saw the brutal murder of his son in a horrifyingly vivid dream, and led police to the killer with an incredibly accurate description.

The murderer of the 18-year-old youth was arrested in his home where police also found large quantities of cocaine.

"We did not believe the father at first," admits police detective Anton Fregosi.

"We were astonished that the victim's father had described the killer perfectly, down to the tiny scar above his lip," he adds.

Detectives had been stumped following the killing of Francis Salerne, out side of Milan, Italy. The body was found near a river with a bullet in the head.

"We knew the boy was unfortunately involved in a drug deal," said Det. Fregosi. "And that made it more difficult to track down."

The young man's father, Luigi Salerne, walked into the police station a week later telling officers he had a vision of the killing in a dream! He told police, "the man who killed my son was in his early 40s, had a receding hair line, a scar above his lip and was missing two fingers on his left hand." Luigi's description fit a known drug dealer. A man named Donnel Feraine. Officers went to Feraine's address where they found the gun believed to be the murder weapon. He was later convicted of the murder. All because of the father's dream!

A psychic dream is a dream in which material from an external source or from an internal source not ordinarily active in the conscious state is received. These dreams contain information from messages, warnings or other communications from individual entities outside the dreamer's consciousness, or they may contain material obtained through psychic abilities of the dreamer, which a dreamer does not normally use in the waking state. Our next dreamer received such a warning, but unfortunately, she did not heed the warning.

Faye Mallett, 34, of Toronto, Canada, was brutally stabbed and killed in the exact spot where she had dreamed she'd seen her own body! Friends and neighbors said that Faye told them about the vivid dream, three days before she was killed.

"She died exactly as she had dreamed, and at the same place, and in the same manner," said Rose Barrett, a shocked neighbor.

Psychic dreams foresees events that will actually occur, as seen in the dream, whereas a warning dream contains an element of urgency, to do something about the potential outcome of a certain situation. Warning dreams portray scenes precisely the way psychic dreams do. The thing about warning dreams is that you are never sure whether or not you can in fact avert the outcome of the dream. However, in a case study, the overwhelming majority of warning dreams present a scene in which the dreamer is an on-looker rather than a participant.

Joseph Ammer's shoe repair shop was located in a run-down section of Indianapolis. On August 7, 1962, the 67-year-old Syrian did not come home for lunch as usual. His wife had just awakened from a terrible nightmare. She told police later, that in her dream, she had seen her husband struggling with a man who repeatedly hit him with a hammer before running away.

Mrs. Ammer was so upset by the dream that she went to the cobbler's shop, only a few blocks away.

When she arrived at the shop, the door was standing wide open. On the floor behind the counter was the body of Joe Ammer. His hands had been tied behind his back with twine, and he had been brutally beaten to death with a hammer, which lay nearby.

Police reluctantly listened to the distraught woman's description of the killer she had seen in her dream a few hours before. Oddly, they got a tip that a man answering the description given by Mrs. Ammer had been seen washing blood off his hands in a tavern rest-room a few minutes after the time of Ammer's slaying. The suspect was dressed as Mrs. Ammer had described, and also looked like the man in the strange dream.

The trial of William Edmonds took place in May of 1963 in the courthouse at Indianapolis.

The jury found Edmonds guilty as charged and he was duly sentenced to life in prison - for a murder, which the widow had watched in her dream.

Sometimes a warning dream is forgotten until the situation triggers a recall. Such was case in our next dream.

The person had a particularly vivid dream. He saw himself in his car driving along a country road, which he did not recognize. In his dream there appeared ahead of him the headlights of an oncoming car, traveling at great speed in his direction. The person did not actually see the crash, but felt that a crash was imminent as the dream ended. Nevertheless, the dream slipped his memory, as time went by. One night he found himself driving along unfamiliar roads. As he was about to round a bend in the road, he noticed in the distance the headlights of an on coming car. At that precise moment, the road seemed suddenly familiar to him. In an instant he recognized it as the strange road in his dream, and the coming car as the headlights of his dream car. The recognition of this situation allowed him to take evasive action, just in time to avoid a head on collision.

Prophetic dreams are dreams in which some element, some information, pertaining to the future, is received. What all prophetic dreams have in common is the element of future events that have not yet transpired, that have not yet begun to shape up in any form whatever, and which therefore could not be foretold by use of the ordinary five senses. Our next dream is just such a dream, and was widely reported by the press.

The worst air accident in U.S. history was the crash of an American Airlines DC-10 only moments after it left the runway at Chicago's O'Hare International Airport in May 1979. Earlier that month the disaster was unreeled in the nightmares of David Booth for 10 haunting nights in a row! Booth had reported his dream - in advance of the actual happening - to the Federal Aviation Administration.

"It was uncanny", said Jack Barker, public affairs officer for the FAA's southern region. He was able to tell us the airline, the type of plane, and description of the crash, including that the air craft would come down inverted, and he told us all of this days in advance of the actual crash. His dream came true."

Similar in its portent to that of Mrs. Ammer was the dream of Mrs. Ernest Topp of East Orange, New Jersey, on May 12, of 1938. Mrs Topp dreamed that she had seen her husband stabbed to death by a small, longhaired man wielding a butcher knife. Mr. Topp was a veteran cook on the steamship *City of Norfolk*, which was en route to London. Three days later she had the same dream. When she awoke screaming - she noticed that it was 5 a.m.

On the afternoon of that same day she received a cable from the vessel, which had just docked in London. It notified her that a Spaniard named Christ Magurrio at 5 a.m. that morning had stabbed her husband to death.

ESP dreams are first of all telepathic dreams, in which the dreamer perceives or receives messages or information from another person, either living or dead, but pertaining to the present, even if at a distance.

In Bedford County, Pennsylvania, there stands a stone monument, erected to the memory of a man's dream and an unsolved mystery.

Early on April 24, 1856, Samuel Cox who lived in a little cabin in Spruce Hollow, could not find his two boys, George, seven, and Joseph, five.

Samuel went into the dense woods calling and listening - without response.

By nightfall more than a hundred men and woman had joined the search for the two missing boys.

The search went on for 10 days, with more than 900 people engaged in the hunt for the two missing boys.

Twelve miles from the area of the search, a farmer named Jacob Dibert had a dream.

He dreamed that he was alone in the woods, searching for the missing children. In the dream, he stepped up on a fallen tree, and there before it lay a dead deer. Leaping over the deer he went on down the deer trail and found a child's shoe and beyond that was a beech tree, lying across a stream. Crossing the stream on the fallen tree, he went on over a stony ridge into a ravine through which a small brook trickled; and there, in the by the roots of a great birch tree, lay the missing boys, dead.

Jacob had the same dream three nights in a row, and he told his wife and a friend Harrison Wysong about the dream

Wysong knew where there was such a place, and to ease his friends mind, he took Jacob to the area.

With in five minutes they stepped up on a fallen tree and there lay a dead deer. Beyond it, lay a child's shoe. They crossed a creek on a fallen beech tree. Jacob spotted a giant birch tree, and at the roots of that tree laid the bodies of the children. Dead from exposure, just as Dibert had seen in his dream.

But not all dreams are dreams of disaster and doom, some are more positive. Some have revealed new inventions, inspired great books, foretold great love stories, saved nations, and more. We will deal with a few of these dreams next.

Whether dreams sometimes constitute a preview of the future is a debatable and highly controversial subject. Yet, the evidence is there.

The excavation of the ancient city of Mycenae was the result of a series of dreams by famed archaeologist Heinrich Schliemann.

There is also the well-documented case of James Watt, inventor of the steam engine. Watt had the same dream night after night for seven days. In his dream, he seemed to be walking through a heavy rainstorm, in which the rain itself turned into tiny lead pellets.

Watt melted some lead over a small fire, he dumped a kettle full into the water filled moat far below. He found that the lead had formed into round balls ideal for loading into guns. And from that day to this, dropping molten lead into water - thanks to a dream, has made lead shot!

All kinds of people have "true dreams," that is to say, dreams that later come true. The ability spans every conceivable class of people, and there is absolutely no way of narrowing it down to any specific group of individuals. Our next dream came true, and it was a very rewarding dream at that.

Several nights in a row Pearl Anderson dreamed that buckets of money were pouring out of a slot machine for her. So she went to Reno and wound up winning the million-dollar jackpot!

Pearl walked into Reno's Club Cal Neva early in the morning, and went to the millionaire slot machine. It took her just two pulls of the handle and $6 to hit the win of a lifetime!

"This just shows you can't ever ignore your dreams, because they can lead to the pot of gold at the end of the rainbow." Said the Oakland, Calif., nurse's aide."

"My dream was short and to the point," said Pearl. "In the dream, I stood in front of a slot machine. All of a sudden the slot machine started gushing silver dollars - and I knew that they were for me."

Pearl's husband Ralph, said; "Pearl woke up several mornings before she hit the big jack pot, and the first thing she said was, 'I dreamed I won a million dollars!' She told me about this crazy dream that had buckets of money gushing from a slot machine. She said, 'I think this could be a lucky sign!' I said, 'If you think so, then you better go to Reno!'

She only had $150 in her purse, but got into her car, and headed east toward Reno. And the rest is history!

Precognitive dreams are the most common of all inexplicable mental experiences. The wonderland of the human mind contains many secrets, which defy research. Lets look at a few.

On November 27, 1952, Paolo Grillo did not go to work in his fields. Instead, he went about all morning telling his neighbors about a strange dream he had the night before.

"Last night," he said, "I dreamed of a very good friend, who died in my arms during the war. He came and stood beside me, and said, "be ready Grillo, today you will come with me."

Grillo went home and sat down in his old chair, and that was where they found him at sundown - dead from a heart attack.

Science itself has profited greatly from the dreams of great men, and of men who became great.

Dr. Herman Hilprecht, famed Assyriologist at the University of Pennsylvania, was writing a book entitled "Old Babylonian Inscriptions". Two small agate fragments and the inscriptions upon them were holding up the completion of the book.

Dr. Hilprecht in a vivid dream saw a tall, thin priest of ancient Babylon standing before him, smiling. In a soft clear voice, the priest said in English: "Come with me. I will help you."

Dr. Hilprecht arose to follow his strange visitors. He led the way down a hall into a room containing a heavy chest containing scraps of agate.

"The two fragments of agate which you have listed as separate articles really belong together, for they are part of an inscribed votive cylinder of agate sent by King Kurigalzu to the temple. It was cut into three parts, in this very room, one of those parts has been destroyed."

"Can you tell me the original inscription?" asked Dr. Hilprecht.

The answer from the priest was, "To the god Ninib, son of Bel, his lord, has Kurigalzu, Pontifex of Bel, presented this."

The priest in Dr. Hilprecht's dream had revealed the true location of the missing treasure room. He had correctly identified the two agate fragments as part of a single cylinder instead of separate objects, a fact later confirmed, and he corrected the translation of the word "Nebuchadnezzar" into the form now accepted by scholars.

The priest had solved these riddles after thirty centuries by Dr. Hilprecht's dream!

Upon rare occasions, dreams have gratifying endings, as in the case of Julius Dittman, and the dream he had in April of 1956.

Mr. Dittman's store was on Huron Road and Ontario Street, in Cleveland, Ohio.

He had a dream that the parking garage had suddenly collapsed on his store, crushing it into rubble. The dream was so vivid that the next morning Mr. Dittman took out a $120,000 insurance policy, which was to take effect April 6, 1956.

On April 7, the parking garage did in fact collapse, and the insurance company had to pay.

There are some dreams that seem to project years, and even decades into the future as in the case of Adrian Christian.

In the year 1833 he dreamed that he was Captain of a ship. His brother, Thomas, was on another ship that was sinking and he rescued him. Adrian Christian told the story to his family, and the mother wrote it into the flyleaf of the family Bible. Adrian had the same dream five times over the next three years, forty-seven years later, Sea Captain Adrian Christian sighted a sinking vessel and rescued 269 persons, including his brother Thomas.

From time immemorial men have been baffled, and amazed, by the mental phenomena called dreams. Sometimes frightening, sometimes confusing, dreams occasionally have still another strange quality - as Captain Christian discovered - they seem to transcend both time and space.

Dreams seem to enable the recipient to ignore time and distance with results, which sometimes surpass the concepts of conventional science.

In Malaysia's mountainous jungles, the Senoi people let dreams dictate the timing and conduct of every important event in Senoi life, and the interpretations of dreams is the major business of the tribe.

Abraham Lincoln dreamed of his death only a few days before John Wilkes Booth murdered him. As Lincoln described the dream to his wife, he was walking through the White House and saw a body laid out. Lincoln asked who the dead man was. "The President," came the answer. "He was killed by an assassin."

The sleeping mind, especially during dreams, may be uniquely receptive to messages that challenge the boundaries of time and space.

4

Bizarre Human Oddities

Giants in the Land

Nearly all cultures on the earth have their stories of monstrous beings inhabiting some forest, mountain, or cave. The Bible contains many such accounts of giants, which according to Scripture roamed the world in our early history!

We are told in the Bible, for example, that Og, King of Basham, was a giant who was almost 17 feet tall.

Perhaps, the best-known giant in the bible is Goliath, the one David killed with a slingshot. He is reported to have been over 10 feet tall, and wore a coat of scale armor of bronze weighing over 125 pounds. Goliath had four brothers that were also giants, according to the testimony of the Bible.

On January 6, 1973, Frank Tolbert, a journalist for the Dallas morning news, reported, "What may be the clearest of giant man tracks yet discovered."

The footprints measure almost 22 inches in length. Dr C. N. Dougherty of Glen Rose, Texas, stated "near the fossilized human footprints are also the deeply engraved prints of three-toed dinosaurs."

European history also contains many accounts of giants. For example, the body of Orestes, according to the Greeks, was almost 12 feet tall.

Pliny describes Gabbaras, who was brought to Rome by Claudius Caesar from Arabia as 10 feet tall, and adds that the remains of Posio and Secundilla, found in the reign of Augustus Caesar in the Sallustian gardens, measured 10 feet 3 inches each.

Herodotus says that the shoes of Perseus measured about 3 feet. Josephus tells of Eleazar, a Jew, among the hostages sent by King of Persia to Rome, who was 11 feet tall. Saxo, the grammarian, mentions a giant almost 14 feet tall, and said he had 12 companions who were much bigger. Ferragus, slain by Roland, the nephew of Charlemagne, was 11 feet tall.

In the twelfth century under the rule of King Eugene II of Scotland, lived a man almost 12 feet tall, according to historic records. The Emperor Maximilian was 9 feet tall, and used his wife's bracelet for a ring. He ate "40 pounds of flesh a day, and drank 6 gallons of wine."

South America also had its giants. The naturalist Turner says that on the river Plata near the Brazilian coast he saw naked savages 12 feet high. Turner confirms this by saying that on the coast of Africa he saw on a boat the body of a 12-foot giant. He claims to have measured the remains himself.

In his account of a voyage to the Straits of Magellan, Jacob Lemaire says that on December 17, 1615, he found at Port Desire several graves covered with rocks, and beneath the stones were skeletons of men, which measured between 10 and 12 feet tall.

The celebrated anatomist Riola, says that there is to be seen in the suburbs of Saint Germain the tomb of the giant Isoret, which skeleton measures 20 feet.

Perhaps the bones of the biggest man ever found was the skeleton of a giant found near Mezarino in Sicily in 1516, whose height was at least 30 feet. Each tooth weighed 5 ounces; and in 1548 there were others found of the height of 30 feet. In Bohemia in 758 it is recorded that there was found a human skeleton 26 feet tall, and the leg bones are still kept in a medieval castle in that country.

Dr. Adam Clark, who died in 1832, measured a man almost 9 feet tall. Early in 1837 there was exhibited at Parma a young man who was 9 feet tall, and weighed 400 pounds. Robert Hale who died in Yarmouth in 1843 was 7 feet 6 inches high and weighed 452 pounds.

The New York Times on December 2, 1930, carried an item that told of the discovery of the remains of an apparent race of giants who once lived at Sayopa, Sonora, a mining town 300 miles south of the Mexican border. A mining engineer, J. E. Coker, said that laborers clearing ranch land near the Yazui river "dug into an old cemetery where bodies of men eight to ten feet in height, were found buried."

On February 19, 1936, the New York Times ran a piece date lined Managua, Nicaragua, which stated that the skeleton of a gigantic man, with the head missing, had been unearthed at El Boquin, on the Mico river, in the Contales district. "The ribs are a yard long and four inches wide and the shin bone is too heavy for one man to carry."

Bizarre Human Oddities

On June 17, 1871, there were married in London, Captain Martin Van Buren Bates, familiarly known as the "Kentucky Giant," and Miss Anna Swann of Nova Scotia, both of whom were over 7 feet. The combined height of these two were greater than that of any couple known to history. His body was well proportioned and his weight was 460 pounds. The husband and wife team traveled as a curiosity from 1866 to 1880, they visited the United States, Canada, Great Britain, France, Spain, Germany, Switzerland, Austria, and Russia.

The copy of the Genesis Apocrypha discovered at Qumran dates back to the 2nd century B.C., but it was obviously based on much older sources. (See Secrets from The Caves in this book.) When discovered in 1947, it had been much mutilated from the ravages of time and humidity. The sheets had become so badly stuck together that years passed before the text was deciphered and made known. When scholars finally made public its contents, the document confirmed that celestial beings from the skies had landed on planet earth. More than that, it told how these beings had mated with earth woman and had begat giants.

In Genesis 6:1-4 in the Old Testament Bible, "sons of God" were captivated by the beauty of the "daughters of man." They subsequently marry them and produce an offspring of giants known as the Nephilim. Genesis goes on to say that these Nephilim were "mighty men and men of renown."

The term "sons of God" is used four times in the Old Testament, each time referring to angels!

In Job 1:6 and Job 2:1, the "sons of God" came to present themselves before the Lord in Heaven. Among the sons of God was Satan.

Since the designation "sons of God" is consistently used in the Old Testament for angels, it is logical to conclude that the term in Genesis 6 refers to fallen angels.

Two New Testament passages shed further light on Genesis 6. They are Jude 6-7 and 2 Peter 2:4. These verses indicate that at some point in time a number of angels fell from their pristine state and proceeded to commit a sexual sin with the daughters of man. Jude 6-7 states:

"And the angels which kept not their first estate, but left their own habitation, he hath reserved in everlasting chains under darkness unto judgment of the great day. Even as Sodom and Gomorrah and the cities about them in like manner, giving themselves over to fornication, and going after strange flesh."

The angels had abandoned their own domain and invaded a realm that was divinely forbidden to them—a realm of created beings of a different nature.

Genesis 6:4 states; "There were giants in the earth in those days." When the word Nephilim is used in Numbers 13:13, the question of size and strength is explicit. Here we are left in no doubt as to their superhuman prowess. When Joshua's spies reported back from Canaan, they called certain of the inhabitants of Canaan "giants." "And there we saw the Nephilim, the sons of Anak, which come of the Nephilim, and we were in our own sight as grasshoppers, and so we were in their sight."

Of all the imaginable phenomena on earth, the progeny of the union between extraterrestrial (the fallen angels) and humans is the most bizarre.

For one thing, they were giants in size and strength. Much documentation of the exceptional physical stature and super-human strength of the Nephilim exists, and this is not the least surprising, knowing that they were "fathered" by angels. According to (Psalms 103:20) angels "excel in strength" and according to (2 peter 2:11) are "greater in power and might" than the men of earth

In Genesis 6, where the word "Nephilim" is first used, we are told that the Nephilim appeared on the earth just before the flood, and their appearance was the main reason for the flood.

The second eruption was probably on a more limited and restricted scale than the first. This was in the land of Canaan, and explains why the Lord ordered their complete destruction. However, Israel, as so often in her history, failed to obey God, and there is reason to believe that some of the Nephilim survived.

The progeny of these Nephilim went under various names. All shared the characteristics of being huge, tall, and strong. Here is an Old Testament description of the Emim, "which were also accounted giants."

Flavius Josephus, the noted Jewish historian of the first century A.D., described these giants as having "bodies so large and countenances so entirely different from other men that they were surprising to the sight and terrible to the hearing." And he adds that in his day, the bones of many giants were on display!

These giants were not confined to the Middle East, two- dozen human footprints of abnormal size have been found in the Paluxy riverbed in Texas. Some of these measured eighteen inches long. Other giant markings have been discovered in such diverse places as Colorado, New Mexico, Arizona and California.

Similar giant footprints have been discovered in other countries especially in the Mt. Victoria region of Australia.

In 1936 Larson Kohl, found the actual skeletons of giants. The bones of a gigantic 18-foot man were found on the shore of Lake Elyasi in Central Africa. Other giant skeletons have been found in Hava, the Transvaal, and China. The evidence for the existence of giants is incontrovertible. "A scientifically assured fact," says Dr. Louis Burkhalter.

Dwarfs & The Littlest People

In the seventeenth century Van Helmont stated that there were pygmies in the Canary Islands, and Abyssinia, Brazil, and Japan. Relics of what must have been a pygmy race have been found in the Hebrides, and in this country in Kentucky and Tennessee. Little people are also a reality.

Dr. Schweinfurth, the distinguished African traveler, confirms that there was a race of pygmies near the source of the Nile river, who were between 3 and 4 feet tall. Another pygmy people near the Obongo, called the Kenkobs are also between 3 and 4 feet high.

In 1869, two full-grown pygmies in Africa measured 33 inches, and weighed less than 20 pounds.

At one time a dwarf was a necessary appendage of every noble family. The Roman Emperors all had their dwarfs. Julia, the niece of Augustus, had a couple of dwarfs, Conopas and Andromeda, each of whom was 2 feet 4 inches in height.

Alypius of Alexandria, was only 2 feet high, and was noted for his wit and charm. Lady Montague describes her dwarfs at the Viennese Court as "devils bedaubed with diamonds." They had succeeded the court Jester and exercised some parts of this ancient office. The court dwarfs were allowed unlimited freedom and speech, and in order to get at truths other men were afraid to utter one of the Kings of Denmark made one of his dwarfs Prime Minister.

Sometimes many dwarfs were present at great noble gatherings. In Rome in 1566, the Cardinal Vitelli gave a sumptuous banquet at which the table attendants were 34 dwarfs. In 1710, Peter the Great of

Russia gave a celebration in honor of the marriage of two of his favorite dwarfs, and 72 dwarfs of both sexes were present at the bridal party.

In England and Spain the nobles had the portraits of their dwarfs painted by the celebrated artists of the day.

Several famous people in history have been dwarfs, men of extraordinary intelligence, including; Attila, the historian, Procopius, Gregory of Tours, Pepin Le Bref, Charles, King of Naples, and Albert the Grand. Godeau, made such a success in literature that he was named the Archbishop of Grasse. The Dutch painter Doos, and the English painter Gibson were both a little over 2 feet tall. Prince Eugene, and the Spanish Admiral Gravina were also dwarfs.

Geoffrey Hudson, from England, was only 18 inches high when he was 30 years old. He proved his bravery and allegiance by assuming command of a royalist company.

In 1730, Wybrand Lokes, a very skillful jeweler, married a woman of normal size, although he was just 25 inches tall. They had four children, all of normal size.

In recent years, several dwarfs have commanded the popular attention. Charles Stratton surnamed "Tom Thumb", was born at Bridgeport Conn., on January 11, 1832; he was above the normal weight of the newborn. However, he quit growing at about five months, when his height was less than 21 Inches. Barnum had him shown all over the world. He was presented to Queen Victoria, and the Royal Family in France. He amassed a fortune, and retired in America.

The Rossow brothers have been recently exhibited to the public. These brothers, Franz and Carl are 20 and 18 years respectively. Franz is said to weigh 24 pounds and measure 21 inches in height; Carl is said to weigh less than his brother, but is 27 inches in height.

The "Lilliputians", were a troupe of singers composed entirely of dwarfs; they appeared in most of the large cities in the United States, and were considered a class act in burlesque.

The "Princess Topaze" was born near Paris in 1879. She was perfectly formed and very intelligent. She was 23 inches tall, and weighed only 14 pounds at maturity. Her parents were both of normal stature,

Miss H. Moritz, an American dwarf weighed 36 pounds and was only 22 inches tall at the age of twenty. Her exhibitor made a fortune with her, and she was among the highest paid in modern times.

A Dutch boy, Phil Bloton quit growing at the age of six months. At that time he was 12 pounds; by the time he was 15 years old, he had gained only 4 pounds. He stood 16 inches tall, and is said to be the world's smallest dwarf. He is still alive today, and makes his living as a "freak" in a sideshow, and has amassed a small fortune billed as "The smallest man alive."

Idiot Savants & Mental Marvels

Jacques Inaudi, earned his living by begging and solving the most difficult problems in arithmetic. In the presence of the Anthropological Society in Paris, in 1880, he was given verbally a task in multiplication, composed of some trillions to be multiplied by billions. In the presence of all the members, he accomplished his task in a few minutes, without the aid of pencil and paper, solving the whole problem in his head. He could perhaps surpass anyone in the world in this particular feat. One by one the learned men would give him complicated math problems, and the lad, would have the answer within seconds. Some times he would give the answer before the question was even completed. He said that he proceeded from left to right in his calculations, instead of from right to left in the usual manner.

Oscar Moore was born blind in Waco, Texas in 1885. It was observed that after touching a person only once with his fingers, he could there after unfailingly recognize and call by name the one whose hand he again felt. Even if months or years had gone by. But his memory was his most remarkable peculiarity. From two years old, and after, he could recite verbatim, all he had ever heard! He has mastered an appalling array of statistics, such as the areas in square miles of hundreds of countries, the population of the world's principal cities, the birthdays of all the Presidents, the names of all the cities of the United States. He remembers the words to every song he has ever heard, and can recite them word for word, after hearing them only once.

William Kolor, could remember the day that every person had been buried in the grave yard where he worked. Going back some 40 years, he could repeat with unvarying accuracy, the name and age of the deceased, and the names of every mourner that attended that particular funeral. But other wise he was a complete fool.

Tom Hicks could repeat all that he had ever written, and whole books that he had read. He could recite them word for word, including the Bible and other such voluminous works. Themistocles could call by name the 20,000 citizens of Athens. Cyrus could call the name of every man in his army.

Rube Fields of Johnson County, Missouri, is a mathematical wonder. His feats have been remembered for twenty years by his neighbors. "Give Rube Fields the distance by rail between any two points, and the dimensions of a car wheel, and as soon as the statement has left your lips, he will tell you the number of revolutions the wheel will make in traveling over the track. Call four or five or any number of columns of figures down a page, and when you have reached the bottom he will announce the sum. And this he will do all day long, without apparent effort or fatigue. He calls his answers quick and sharp, seemingly by intuition. The size of the computations seem to offer no bar to their rapid solution, and answers in which long lines of figures are reeled off come with perfect ease." In reviewing the peculiar endowment of Fields, the Chicago Record says:

"How this feat is performed is as much a mystery as the process by which he solves a problem in arithmetic. He answers no questions. Rapid mathematicians, men of study, who by intense application and short methods have become expert, have sought to probe these two mysteries, but without results. Indeed, the man's intelligence is of so low an order as to prevent him from aiding those who seek to know. With age, too, he grows surlier. Of what vast value this "gift" might be to the world of science, if coupled with average intelligence, is readily imagined. That it will ever be understood is unlikely. As it is, the power staggers belief and makes modern psychology, with its study of brain cells, stand aghast. As to poor Fields himself, he excites only sympathy. Homeless, unkempt, and uncouth, traveling aimlessly on a journey, which he does not understand, he hugs to his heart a marvelous power, which he declares to be a gift from God. To his weak mind it lifts him above his fellow men, and yet it is totally useless to the world."

Ivan Birch, can tell you what the weather was like every day for the last 59 years! With out failure, or second thought, he can site the high and low temperature, clear sky, or overcast, rain or shine, and the exact time of sunrise and sunset. Yet, he has a hard time feeding himself. Dr. Michael Coleman of the United States Weather Service, ran a computer print out for the time of sunset and sunrise for a period of ten years some forty years ago in order to check his memory. Dr. Coleman was "astonished beyond description" when Ivan reeled off the numbers, while he was watching television! "After some five hours, he made a believer out of me," Dr. Coleman said. "How he can remember those kind of numbers is more than I can understand." Birch also gave Dr. Coleman the wind velocity, and other bits of information, such as the direction it blew.

Bill Mong had a memory for one thing - he was a walking encyclopedia! I mean all thirty volumes that were in the library of the house where Bill was raised. He had memorized -word for word- the complete thirty volumes.

You could ask him any question about the information contained in any of the books, and without hesitation, or error, he would recite verbatim the information from his memory.

Dr. J. M. Walker interviewed Bill for one whole afternoon. "I was astonished " said Dr. Walker, "I could even ask Mr. Mong what was on a certain page, in a certain volume, and he would quote it word for word!"

"At one point I asked Bill to quote from the top of page 106, volume 6, and without any delay - he began. I watched him in total disbelief for over an hour, as he continued. Page after page, and he never missed a word!"

"Yet, Bill Mong can't figure out how to tie his own shoes, he has never had a job, and at 41 years of age, it is doubtful that he ever will."

Bill still lives at home with his dad Roger, who must provide for all of his needs.

"It is doubtful that Bill understands all that he has memorized," said Dr. Walker. "It one point I asked him what a certain passage meant, and he just repeated the passage over, and over again."

Somewhere in this complex remarkable endowment lies an explanation - but thus far no scientist has been able to explain it. However, the subconscious mind forgets little or nothing; it remains only to develop the ability to recall from the subconscious to be able to perform amazing feats of memory.

Perhaps the best-known case of such fabulous ability to recall was the Rabbi Elijah of Vilna, a Lithuanian, who regarded his strange mental power as a curse. At will he could recall any page and any portion of any page from any of the thousands of books that he had read! He was unable to forget anything that he had read. He said; "It was like living in a library."

Leon Gambetta, could repeat verbatim the complete works of Victor Hugo, which were many thousands of pages. He could repeat the works forward or backward, beginning at any desired point in any volume, word for word.

Harry Nelson Pillsbury, the American chess wizard, could play twenty games at once, and recall more than a thousand moves that had been made, and play a game of cards at the same time.

Mathurin Veyssiere, had an incredible memory for sound. After hearing a speech in any language he could repeat it with proper pronunciation and the original accent.

Zerah Colburn could solve involved mathematical problems in his mind. During one exhibition, he was asked to multiply 21,735 by 543, and the boy had the correct answer immediately. When he was asked how he got the answer so quickly, he promptly replied: "Oh that was easy - I just multiplied 65,205 by 181!"

Johann Martin Dase, born in Hamburg in 1824, was so fast and reliable with complex computations that scientists of his day often asked him to solve involved astronomical problems - like 100 digits multiplied by 100 digits. This he did in a matter of a few minutes in his head.

The human mind is a wonderland of incredible capabilities, little understood, and largely unexplained.

Tom Fuller was an illiterate seventy year-old slave - but he had the ability to outperform such modern marvels as the computer. A schoolteacher went to the plantation where old Tom worked to put him to the test, and came away mystified. The first question was to reduce his 70 years, 12 days, and 12 hours into seconds. In 30 seconds Tom had the answer. (Which took into account the leap years involved.)

The owners permitted scholars to come and test Tom. Without exception, he answered all questions in a matter of seconds! Yet, Tom had never been to school, and could not read or write.

Another man whose strange powers baffled science was Charles Cansler of Knoxville. Cansler toured the country giving demonstrations of his remarkable abilities. He could carry in his mind figures involving many millions, and could instantly give the squares of such numbers, with a speed that beat the best existing calculators.

In his customary performance, he would turn his back to a blackboard, while spectators would write long columns of figures on the board. Then he would walk to the board and instantly write the correct total of the figures on the board.

Jebidiah Buxton was born in England in 1707. He had no schooling, but could also do complex calculations. In one test he calculated how much a farthing would be worth if it were doubled 139 times. He gave the answer in 39 digits instantly. An oddity of Buxton's performance was that he could pursue his normal activities, whatever he might be doing at the time, while his brain carried on its complex functions unassisted.

Once, while he was intently watching a dog fight, he solved a problem involving 42 digits in 10 seconds!

Natures Footprints

Sir Francis Galton, in his long search for statistical evidence to prove the inheritance of mental ability found that the life history of twins resulted in a vast amount of information and insight. The results of his research was published in *Fraser's Magazine* in 1875 under the title *The History of Twins, As A Criterion of the Relative Powers of Nature and Nurture.*

The following excerpts highlight his scientific insight:

My materials were obtained by sending circulars of inquiry to persons who were either twins themselves or the near relatives of twins.

I have received about eighty returns of cases of close similarity, thirty-five of which entered into many instructive details. In a few of these, not a single point of difference could be specified. In the remainder, the color of the hair and eyes were almost always identical; the height, weight, and strength were generally very nearly so. In some cases, not even the twins themselves could distinguish their own notes of lectures, etc., many cases in which the handwriting was indistinguishable.

The impression that all this evidence leaves on my mind is one of some wonder whether nurture can do anything at all beyond giving instruction and professional training.

There is no escape from the conclusion that nature prevails enormously over nurture when the differences of nurture do not exceed what is commonly to be found among persons of the same rank of society and in the same country.

It seems contrary to all experience that nurture should go for so little.

No new study of importance was published until E. L. Thorndike's celebrated monograph appeared in 1905. Thorndike added structured psychological tests to his study of twins.

Although conducted 25 years apart, and on two different continents, the results of the first two methodical twin studies are remarkably similar. Thorndikes sums up his conclusions:

Doubtless we all feel a repugnance to assigning so little efficacy to environmental forces as the facts of this study seem to demand; but common opinion also feels a repugnance to believing that the mental resemblance of twins, however caused, are as great as the physical resemblance. Yet they are!

Curtis Merriman in 1924 was the first to employ standardized individual and group IQ tests to test the intellectual similarities of twins. The results of his investigations finally convinced psychologists that "environment appears to have no important influence on the degree of twin resemblance in mental capability."

This resemblance also extends to ability, personality, and interest!

Without attempting to interpret these correlations precisely at this point, their most obvious implications are that individual differences in all traits of behavior, from general intelligence to fingernail biting, are due mostly to genetic factors! The following story is a good example:

Twin brothers Ronnie and Donnie Shaw proposed to their twin-sister wives Dawn and Shaune on the very same night, unaware of each other's proposal - and even more mind-boggling coincidence, the women had babies within 3 hours of each other on June 23, in the same hospital! But those are just two of the astounding parallels in the lives of the twin couples.

They married on the same day, honeymooned together - and now live just four houses apart in Arlington, Texas.

For two years the two couples dated. Ronnie and Donnie did not know the other was planning on proposing on the same night. The two were off on a date by themselves. After they arrived home, Dawn said to Shaune, "guess what - we're getting married!" No sooner had Dawn said these words than Shaune replied, "well, guess what, we're getting married, too!"

So they had a double wedding. On June 14, 1980, the twin brothers married the twin sisters in a double ceremony.

They honeymooned together in New Orleans, and then later moved into homes near each other. Soon both became pregnant.

Oddly enough, both wives gave birth two days before their babies were due - and at first, neither knew the other was even at the hospital!

There are many strange parallels in the lives of the twins; the brothers suffer from nosebleeds - and for some bizarre reason they always get them on the same day!

The girls frequently go shopping separately, and come home with the exact same outfits!

They even sent their mother the very same birthday card - even though they bought them separately, and that's happened on more than one occasion.

It became clear from a test involving 750 sets of twins, that we inherit our IQ, and most of our personality traits including: irritability, practicality, punctuality, ambitiousness, persistence, nervousness, submissiveness, impulsiveness, emotional control, and seriousness.

We also inherit many of our so-called likes and dislikes; our talents - such as music, art, and sports ability, to name just a few, and even our taste for foods, job preference, and clothes!

A good example of this is a set of male twins named Donald and Ronald. They were separated at birth, and adopted into different homes more than 1,000 miles apart.

A lower middle class couple adopted Donald, while an upper middle class family adopted Ronald.

For many years they were unaware that they had a twin brother. However, due to this study that was investigating twins who were raised apart, they finally became aware of each other.

At age 35 they were re-united for the first time, and the similarities were both astounding and far-reaching in their implications.

Both were married to a girl with the same first name, both had three children, of which two had either the same first, or the same middle name. Both were the managers of a fast food restaurant, and both had a small dog by the unusual name of "Tad."

When questioned about how the name for the dog was selected, they both indicated that the name "just came to them."

They both bit their fingernails, and had a nervous twitch when they spoke to strangers.

They both said that they "always wanted to work in the restaurant business," and both started at the bottom, right after graduation from high school, and worked their way to manager in three years. Neither wanted to attend college as a "personal choice."

They both wore clothes that were very similar, and when they met for the first time, they both were wearing jeans, a sport shirt, and loafers. "It was almost as if we were looking into a mirror," said Ronald.

It was also noted that both parted their hair in a similar manner.

They both expressed a feeling that something was missing in their lives - until they were brought together.

In a similar case, twins Larry and Jerry - met for the first time at age 25. They were married to girls with the first name of Mary and Terry. Both had one boy. Their first and middle names were William Robert and Robert William. (It has been noted that when an analysis is done on a family tree, the same names keep popping up, regardless if one is aware of those names or not.)

Larry and Jerry were both in the performing arts, and both played several different musical instruments.

When the wives asked the twins, what they would like to eat, their first night together, they both answered at the same time and said, "steak and ale."

Brain wave testing was done on 200 sets of twins, which established the fact that brain waves are an inheritable trait.

Another such investigation in 1965 by Dustman & Beck concluded: "this fact cannot be questioned." Both the right and left hemisphere of many of these twins responded almost identically to many different types of stimulation. This Brain-wave tracing started in the waking state, and continued into sleep, and dreaming. The patterns were almost the same, and even dream content was similar!

In spite of the many weaknesses in the twin method, human twins provide the best subjects available for the study of genetic-environmental interactions.

The idea that twins can be used as a natural experiment to determine behavior is an old one. St. Augustine of Hippo, in his masterpiece *The City of God*, cited evidence from twins in order to refute astrology and support Christianity. The Bible states that even the hairs on your head are all numbered, and that the sins of the parents are passed on to the third and forth generations. (A fact not known in scientific circles until almost two centuries later.) The Bible also asks the question: "Can a leopard change it's spots?"

Francis Galton, however, was the first person to frame the problem in a scientific context. In 1875 he wrote:

The exceedingly close resemblance attributed to twins has been the subject of many novels and plays and most persons have felt a desire to know upon what basis of truth those works of fiction may rest. But twins have many claims to attention, one of which is that their history affords a means of distinguishing between the effects of tendencies received at birth, and those that were imposed by the circumstances of their after lives; in other words, between the effects of nature and of nurture. The objection to statistical evidence in proof of its inheritance has always been: "The persons whom you compare may have lived under similar social conditions and have had similar advantages of education." But such prominent conditions are only a small part of those that determine the future of each man's life. It is to no trifling accidental circumstances that the bent of his disposition and his success are mainly due."

Sandra Scarr recently summarized her views regarding the influence of heredity and environment on behavioral development and, without citing Galton, came to a conclusion remarkably similar to his:

Ordinary differences between families have little effect on children's development, unless the family is outside of the normal developmental range. Good enough, ordinary parents probably have the same effects on their children's development as culturally defined super-parents. Children's outcome do not depend on whether parents take children to the ball game or to a museum so much as they depend on genetic transmission, on plentiful opportunities, and on having a good enough environment that supports children's development to become themselves."

Without doubt, twins can be usefully used as a tool for addressing a wide variety of interesting questions regarding the genesis of human traits, including behavioral factors.

After IQ, twin researchers have studied personality more frequently than any other trait.

Lay friends or relatives have asked virtually every behavioral geneticist whether schizophrenia, depression, or even alcoholism is genetic. T. J. Bouchard, Jr., and P. Propping, in their book *Twins As A Tool Of Behavioral Genetics,* in response to this question write: "There is strong evidence for genetic influence."

Singer and Berg (1991) write; "Genes and genomes constitute the blueprints of life. They constitute the machinery of gene replication and gene expression."

D. W. Fulker and L. R. Cardon of the *Institute for Behavioral Genetics* write: "Twin study is arguably the single most powerful method in human quantitative genetics. It is easy to carry out, since twins are readily available in the general population. It is statistically powerful, it is straightforward to analyze, and its outcome is capable of yielding a variety of insights into the structural properties of traits. When combined with data on parents and offspring, adoptions, and inbreeding, we can validate the basic twin approach and fill out the picture in terms of the biological and social significance of the trait in question."

"We have found that the cumulative data suggest a mode of inheritance for general intelligence."

Sex differences in cognitive abilities have been recognized for many decades. Maccoby and Jacklin (1975) concluded that there was a gender difference-favoring girls in verbal ability and boys in quantitative and spatial abilities.

The study of twins also provides a way to identify specific brain regions that can be the focus of increasing detailed study in psychiatric populations.

The Genain quadruplets, Nora, Iris, Myra, and Hester, were reared together by their parents. By age 24, all four had received the diagnosis of schizophrenia. They all became ill within a few months of each other.

A study by Begleiter and Kissin in 1993 concluded in brief, that alcoholism shows a strong tendency to run in families. Adopted away children of alcoholics still have a strong tendency to alcoholism. Alcoholism in adoptive parents, in contrast, is not a predictive of alcoholism in their adopted children. The same is true for all substance abuse. Again, I am reminded of the Biblical admonition that the "sins of the parents are passed on to the third and forth generations."

Each life begins with a combination of genes different from any other. Making each of us quite literally, a new experiment of nature. Every one of us is unique. Chemistry and heredity determine eye color, blood type, instill the sex drive, determine the shape of our fingers, and for the most part, the length of our lives.

Grandparents know firsthand of reappearing traits in family bloodlines.

Bizarre Human Oddities

We know now that shyness is genetically influenced. As is many eating disorders. A tendency toward obsessive behavior has its origins in the genes. An ever-widening range of physical and psychological illness can now be traced to genetics. Who would have expected that one of the most humane of all human traits, empathy, had biological underpinnings?

Characteristics, such as chronic ear infections and curiosity, are now understood to be inherited.

Peter B. Neubauer, M.D., and Alexander Neubauer, in their outstanding work *The New Genetics of Personality,* Write:

"These findings have not been easy to accept. They challenge basic theories and at first seem to minimize the cherished power of the environment and our magic wish to mold our children according to our hopes and intentions. Rather than being equally excited by the variations that genes initiate, we still tend to shy away from recognizing them. Rather than esteeming nature's generosity and variations, we fail to observe all the differences that make each life unique from the start. We fail to see that individual makeup at birth colors each child's response to the world in which he grows."

The environment these days, receives too many of the laurels for life's accomplishments and too much of the burden for its failures.

Almost daily new findings are published that confirm the pivotal role genes play in our lives. Genetic engineering has become a practical part of medical intervention. The gene for Alzheimer's and many other diseases have now been isolated. A gene that causes manic-depressive disorders has been found.

As this research precedes, the technology to transplant faulty genes with segments of new, healthy ones may be perfected. Meaning many of the four thousand known inherited diseases (one-fourth of which affect mental functioning) would be cured.

Can parents continue to accept the hereditary transmission of their children's physical traits but fail to accept the transmission of other qualities, such as sensitivity to noise, self-control, or shyness?

Can we continue to believe that children are only who we make them and not also entities unto themselves, with biological pasts and developing futures?

Dr. Neubauer sums it up this way:

"What we bring to the world also influences how we experience it. What we bring - our temperaments, patterns for growth, individual inclinations, and our personal susceptibilities - affects the way we see life and the way we respond to it, from the first moments after birth to the day of our death. Simply put, what we bring is the genes."

Adoptive parents of identical twin girls, who were raised apart, were asked a variety of questions. Commenting on eating habits one mother said, "The girl is impossible. Won't touch anything I give her. No mashed potatoes, no bananas, nothing without cinnamon. Everything has to have cinnamon on it. She wants cinnamon on everything."

In the house of the second twin, far away from the first, no eating problem was mentioned at all by the other mother. "She eats well," she said. "As a matter of fact, as long as I put cinnamon on her food she'll eat anything."

And consider the case of identical men, age thirty, were separated at birth and raised in different countries. Both men were neat to the point of pathology. When the first was asked why he felt the need to be so clean and neat he said: "My mother always kept the house in perfect order. I learned this from her." The man's twin, just as much a perfectionist, explained his behavior this way: "The reason is quite simple. I'm reacting to my mother, who was an absolute slob."

Thomas J. Bouchard at the University of Minnesota, who studied hundreds of pairs of twins, found hereditary links to personality traits such as leadership, vulnerability to stress, traditionalism, and imagination. Shyness was found to be exceptionally heritable. He concludes: "It was once thought that the traits of personality are shaped by the environment over the course of development. We can now say that much of what becomes personality is intrinsic to the child from the start, predisposed by natural inclination."

Twins George and Millan were separated at birth and adopted by different families. The boys met for the first time at age twenty. They were both handsome, athletic young men. Both had won boxing championships. Both were artistic. Physical characteristics were identical, down to the cavities in the teeth. Both developed a progressive crippling disease of the spine within two months of each other.

Some effects of genetic timing are easily recognizable, especially those that shape the body, and those that incline people to specific hardwired illnesses linked directly to the genes.

Alzheimer's disease for example is encoded in the genes at birth but only becomes expressed in late adulthood.

Our inherited susceptibilities and tendencies help determine who we are as individuals, they set the ground plan for our maturation and development.

In order to achieve personal insight, we must each become aware of heredity's role in our lives. We must each discover our inner ranges, potentials, and tendencies. It should simply become one of the goals of growing up.

However, we should not feel that our life is "predestined" for nothing could be farther from the truth. We may be born with certain "built in" personality traits, and be predisposed to certain illness or disease, but that's the price of passage into this world.

Every person is a free moral agent, and we really do have "veto power" over our impulses, and we must be accountable for our actions and reactions.

Many times in life we must go against our "natural inclinations" or resist certain "cravings" because it's the right thing to do. We must learn to cultivate those genetic talents that we are blessed with, and choke out the negative feelings, that would cause harm to others or ourselves. The choice is up to every person who would become a responsible citizen of the world.

5

Fascinating Creatures & Botanical Marvels

Strange Things & Living Creatures

In nature we find some of the strangest things, and living creatures, that one can imagine. Indeed, many of the examples in this chapter, are stranger than you can imagine!

- In the water, the *Australian walking fish* swims like all other fish. But this fish has unusual fins in that they also use for walking. This peculiar fish can stay out of the water for many hours, and it often climbs trees near the water to bask in the sun, and feed on insects that can be found in the trees.
- The squid has a singular way of escaping from its enemies. He shoots out a cloud of black sepia, leaving its enemy in the dark, as he makes his getaway.
- The *bathysphere fish*, on the other hand, has glowing spots along its sides that resemble the portholes of a diving bell. These luminous spots serve to frighten away its enemy, and attract the fish on which it feeds.
- The tiny *hatcher fish* emits from its body ghostly, greenish white lights that resemble a roe of teeth, deterring would be enemies while it feeds undisturbed on its diet of plankton.
- The *lantern fish* uses the blue, green, and yellow luminous "buttons" along its body as a form of recognition and to attract a mate, while the anglerfish actually uses luminous lures as fishing rods to catch fish whose usual prey is luminescent.
- The salmon will swim 1,000 miles against a current to spawn in the precise stream in which it was born. A salmon will mature in the ocean, then return to that river, and swim at speeds up to thirty miles per hour, fighting rapids and leaping waterfalls as high as fifteen feet. The journey does not end until the fish has found the stream in which it was

born. It has been discovered in recent years, that they locate this stream by the sense of smell.

- Many deep-sea creatures also employ bioluminescence, probable both as a mating signal and as a means of attracting prey. In depths where light does not penetrate, sea anemones, sponges, coral shrimps, prawns, and squids can produce their own light. Other fish act as hosts to colonies of luminescent bacteria, which illuminate parts of their bodies from within. The bacteria emit continuous light, but the host sometimes has a means of shutting off the light, presumably when danger threatens. This is done by reducing the oxygen supply.

- Where no glimmer of light pierces the darkness, miles below the surface of the ocean, the anglerfish catches its prey with luminous bait. On top of his head, is a strand of flesh about five inches long. On the tip of that strand is a wad of flesh, which glows in the dark. Smaller fish attracted to the dangling light are quickly eaten.

- The *Indo-Pacific photoblepharon*, a shallow water fish, has a large white spot under each eye, rich in bacteria and blood vessels. A black fold of skin above the luminous spot can be lowered to shut off the light.

- When the time is right, schools of grunions are carried onto shore by breaking waves. The female stands up on her tail and whirls around in a wild dance! As she dances, her tail digs a small hole in the wet sand, in which she deposits her eggs. The male grunion, also swept in by the waves, dances into the holes and fertilizes the eggs. Another wave carries the male and female back into the ocean.

- The *arch-fish* claims its victims by firing a missile. Swimming close to the surface in rivers, it seeks insects on low hanging branches of trees. When it spots a victim, the fish rises to the surface and fires a jet of water from its mouth, knocking the insect into the water.

- The lantern fish is a deep-sea fish that lies at the bottom of the sea and lures its food to it by a light that glows from its mouth.

- The *gigantactis fish* uses a powerful "flashlight" to illuminate its prey. This fish swims at a depth of 6000 feet, and lights its way through the ocean depths by a bright light it carries at the end of a rod projecting from its head.

- Another fish with "running lights" is the *photostomias guerni*. This fish is equipped with two rows of phosphorescent spots to light its way through the depths of the ocean.

- The *acari* fish of South America, which grows to about a foot in length, is completely encased in a bony suit of armor that is transparent. The light from *gigantacis* just shines right through it. The Bay scallop on the other hand has a row of blue eyes that are very sensitive to light, and is blinded by the *gigantacis*.

- The *hemigrammus ocellifer* fish has red eyes that reflect light and a copper spot on its tail that acts as a rear reflector. *Gigantacis* can see this fish coming or going.

- The *mellia tessellata* crab has developed a unique way to defend itself from would be predators. It carries in each claw a *Red Sea anemone* that packs a painful sting. Whenever they get to close, she just reaches out and zaps them with the *anemone*. This crab has learned to use a hand held weapon!

- The frogfish of Asia can live out of the water for many days, and uses its front flippers to walk on land. These front flippers are also used to pick up worms and insects and place them into the mouth of this strange fish.

- The *flying gurnard* swims in water, walks on land, and flies short distances through the air. This *gurnard* fish uses three pectoral projections like fingers to dig food from the muddy bottom of the sea and pop it into its mouth.

- The female *bowfin* fish, for some strange reason, always wants to eat the eggs she has laid. However, the male bowfin fish guards the eggs, and fights off the mother whenever she gets too near.

- The *kurtus gulliver* male fish of New Guinea has a hook on its neck to which the female attaches her newly laid eggs, and the hooked male has to carry the eggs around until they hatch.

- The *butterfly blenny* fish has found a safe place to lay her eggs also. She always lays them in an empty seashell.

- The tiny *hapalocarcinus marsupialis* crab protects itself against enemies on the barrier reef of Australia, by imprisoning itself for life. Soon after birth it takes up its home in the fork of two coral branches, which grow into a cage with openings just large enough to permit entrance of seawater carrying the crabs nourishment.

- The electric eel, sometimes six foot long, can generate electric power strong enough to light a dozen light bulbs! Its electric organs discharge about 600 volts at a current of two amps to kill prey, and to ward off predators.

Fascinating Creatures & Botanical Marvels

- Whereas the electric catfish shocks fish into convulsions - only to eat the food they have swallowed but not yet digested.
- The heart of a shark still beats strongly hours after it has been removed from the body. But the octopus has three hearts, and can live as long as any one of the three works.
- The *marbled sea pearch* fish is the chameleon of the sea. This fish can change its coloring to match its surroundings anywhere in the ocean depths.
- The *parrotfish* of Australia has two sets of teeth. One in its mouth and one in its throat.
- The *upside-down catfish* of South America swim normally for a while when they are young - then they turn over on their backs for the remainder of their lives.
- The *piraruca*, a fresh water fish of Brazil grows to 12 feet in length and some 1200 pounds in weight. It has a tongue so rough that the Indians use it as a grater.
- The *protoperus* fish has lungs. It has no fins or gills, and breaths air by coming to the surface of the water. While the *polypterus* has both lungs and gills. It can breath equally well underwater and from the air at the surface. The African lungfish also has both lungs and gills and walks out of the water on its fins to lie in the sun or to feed.
- The *misgurn* fish is kept under observation by seismological stations in India, because the fish darts rapidly about when an earthquake is impending, and gives several hours warning. The rest of the time the fish normally lies sluggish in the mud-bed of slow streams.
- Amazingly, in cold water the flat-fish turns on it's left side, and the left eye moves to the right side of the head; in warm water, the fish turns on it's right side, and the right eye does the moving! No one knows why temperature should have this effect.
- The most vulnerable creature in all of nature is *ipnops,* a blind deep-sea fish that is helpless to defend itself or elude pursuit and bears on its head a luminous patch that attracts the attention of predators.
- The *cave characin* fish found only in Mexican caves are always blind.
- The largest bird in the world is the ostrich. A bird that can't fly, but can outrun a racehorse. The ostrich can cover twenty-five feet in one stride, and maintain more than sixty miles per hour for long periods of time.

- In birds the sense of sight has also developed to an unusual degree. The high soaring buzzard, which needs to be able to pick out a meal as small as a lizard or even a beetle on the ground below, has eyesight eight times as keen as a man's.

- The homing instincts of the carrier pigeon is so strongly developed that these birds can fly more than 2,000 miles over unfamiliar territory and never fail to find their way home!

- To a sparrow the human voice sounds something like a low rumbling of distant thunder, and the low notes of a singer, are completely inaudible to the bird. This is because the sparrow's hearing range is different from humans.

- As a bat flies through the air, the bat opens his mouth and sends out a series of sounds. If the sound strikes an object, they bounce back, like a radar echo. The radar system of this strange creature is so precise that a totally blind bat can fly about in a crowded room without touching an obstacle with so much as a wing tip. In this way bats can flit about in great numbers without colliding. Even when they fly out of caves in the thousands, they respond only to their own individual echo signals and are not confused by their neighbors' noises. How they manage to tune in so finely remains a mystery. Scientists have conducted experiments in which they tried to jam the radar signals of bats by drowning them with a volume of noise 2,000 times as intense and on the same frequency. It made no difference. Somehow, the bats managed to pick out their own echoes and ignore the foreign noise.

- The *deer botfly*, is a tiny insect, which happens to be the fastest creature in the world. This marvelous mite can fly at the speed of 820 miles per hour! This creature could fly around the world, and never see a sunset.

- The *plover bird* and the crocodile have worked out a mutually satisfactory arrangement, the bird gets food, and the reptile gets service. When the crocodile has finished a meal, he opens his mouth so that the small bird can hop inside and pick the reptile's teeth clean.

- The *dipper bird* has the ability to "fly" under water. Using its strong wings as fins, it dives into the water, and "flies" to the bottom. It can stay down for more than a minute as it searches in the sand for food. A movable flap over the bird's nostrils keeps out the water, and a membrane protects the eyes.

- The *paradise whidah* bird of Africa, always lays its eggs in the nest of a *waxwing* - making certain that the bird has the same plumage and song as the *whidah*.
- The *western king bird* which always dines on bees, attracts them by flashing on its head a brilliant red spot, which bees mistake for a flower.
- The most colorful bird is the *pitta bird* of Australia. It has nine bright colors -blue, green, orange, brown, pink, red, white, purple, and black.
- The noisy *scrub bird* of Australia, which nests on the ground is an excellent ventriloquist, and can imitate the call of any other bird. A predator never knows what kind of a bird is near, or just where the bird is.
- The *ruffed grouse* grows "snowshoes" each year before the winter snows. They develop a fringe of horn on each toe, which prevents the bird from sinking into the snow.
- The *reef heron*, which feeds on shellfish on the great barrier reef, daily flies 30 miles from the mainland, and although the tide changes vary by 45 minutes a day, the heron always arrives at the exact time the water recedes.
- The *shadow birds* always build a 3-room nest. One section is a nursery, the second is a pantry, and in the third the male parent stands guard against intruders.
- The bird that uses a tool. The *woodpecker finch* of the Galapagos Islands in the Pacific has developed the long bill of a woodpecker. But because it does not have the woodpeckers long tongue, it pries insects out of tree crevices with a twig.
- The *song thrush* having captured a snail gets at its meat by smashing the shell against a rock.
- The *jacana bird* of Africa can balance on floating leaves, the bird has the biggest feet of any bird, and are as long as its body.
- The *guillemot* lays a single egg among thousands of others of the same species - but she can always identify her own.
- The *hawk cuckoo bird* smuggles its eggs into the nests of other birds, and lays her eggs in matching colors! She lays brown eggs or blue eggs - always matching the eggs in the nest it has invaded. By this means, other birds raise her offspring.

- The *woodcock bird* never sees its food. Its 3-inch beak is driven deep into the mud and it devours earthworms it has located by sense of feel.
- A two-day-old gazelle can run at sixty miles per hour! Which is faster than a full-grown racehorse. But the strange thing is that this animal never takes a drink of water. Gazelles have a digestive system that extracts water from the food they eat.
- The second largest land animal, the hippopotamus, which can weigh more than three tons, and eats only vegetation, has a very strange feature. When he sweats, his thick, hairless hide exudes an odd carmine red perspiration. Which looks like blood.
- The mobility of the flea depends entirely on their extraordinary leaping ability. The flea can jump one hundred times his or her own height - a feat comparable to a man jumping 400 feet. They have no wings, but jump from one warm blooded animal to another.
- The king cobra is one of the most dangerous creatures in the world. Moreover, this snake will attack with no provocation. Its poison sacs are enormous, and the venom is one of the deadliest known. Even a huge elephant may fall victim to the attack of the king cobra.
- The secrets of the animal world are being gradually revealed. Prof. Bullock, of the University of California, discovered that rattlesnakes possess a "third eye."

When Bullock taped over a rattler's eyes, he found that it could still locate a mouse with uncanny precision. The organs responsible appeared to be two small dimples located on either side of the snake's head, between the nostrils and eyes.

In these two small dimples Bullock discovered heat-sensitive cells that enabled the snake not only to locate living prey when its eyes were covered, or at night, but also to determine the size and shape of these creatures from the heat given off by their bodies.

- The suicidal lemming remains one of the most baffling mysteries of the animal kingdom. The small mouse like creatures that live in the icy northern regions of Scandinavia, frequently commit mass suicide - and no one knows why. After running for weeks the lemmings finally reach the sea shore, and then, row upon row, plunge headlong into the water! The lemming is normally shy and scarce. But lemmings' lives follow a strange four-year cycle. Every four years frenzied millions of these little rodents hurl themselves from Norwegian cliffs and beaches into the sea to drown. Even the smallest waves overwhelm them. Thus they provide their own harsh means of population control!

• The chameleon is a small lizard best known for its ability to change colors. As his mood changes, this remarkable reptile can pass through the colors of the rainbow. But the strangest attribute of this creature is his tongue - which is longer than his body! The chameleon has eyes independent of each other and can simultaneously look in two directions.

• For quickness, few animals can match the otter. He can poke his head out of a hole in the ice, disappear, and pop up through another hole yards away in a split second.

• The *mouse deer* is the image of the full sized deer that roams the forests of America, identical in almost every detail except that the fully-grown mouse deer stands less than one foot high. For his size, the mouse deer is one of the fastest creatures in the jungle. He runs with a peculiar motion, something like the bouncing of a rubber ball, ending every few steps in a leap.

• The *trap-door spider* lives in upright tunnels he digs in the sandy ground. The mouth of the tunnel is covered with a trapdoor that is hinged to the ground on one side. This door is almost impossible to detect when closed. But the spider knows where the tunnels are, and in a split second he can open the door, slip inside, and close the door again, seemingly vanishing into thin air.

• The *goliath beetle* of West Africa has a pair of black horns, each about a quarter of an inch long. The beetle can use these horns to slowly peel a banana.

• The raccoon will eat many things - just as long as it's clean. Before eating anything, the raccoon first washes the food in the nearest available water. He would go hungry rather than eat anything it has not washed first.

• The *kangaroo rat* can jump five feet in the air, and land directly on top of a grasshopper. And when two kangaroo rats fight, they both leap into the air, striking at each other with their sharp claws while in midair! But the most peculiar thing of all about this rat is that he never, in his entire life, takes a drink of water - or any other liquid. His diet of roots and plants provide all the liquid the kangaroo rat needs.

• The *koala bear* is another animal that never drinks water in his entire life, and he'll eat only the leaf of the eucalyptus tree.

- The *giant anteater* has no mouth, or jaws, or teeth. The anteater's snout is actually a tube with a small opening at the end. He catches a few hundred ants at a time on the fly paper like surface of his tongue, and then sucks his tongue back through his tubular snout and swallows the ants whole.
- Occasionally a lion will leap on a porcupine. When attacked, the slow-footed rodent merely turns his back. After one bite the foolhardy cat's mouth is bristling with sharp quills. The tip of each quill is barbed, the more the lion struggles to spit out the quills, the deeper they stick into his throat. In a few days, unable to eat, the lion will die.
- The shrew is only two inches long, and weighs about as much as a nickel. The shrew has a bite like a cobra, eats three times its weight in meat every day, and can whip an animal three times his size.
- The turtle lives longer than any other creature on Earth. Some turtles live for more than 200 years. The turtle may be one of the slowest creatures in the animal kingdom, but he makes up for it in longevity.
- Animals that hunt by following a trail often have poor eyesight but a keen sense of smell. Dogs see a world that is blurred and devoid of color. But a dog's sense of smell is one million times better than mans.
- There is a type of beetle larva that feeds on vine roots and is sensitive only to a single smell - that of carbon dioxide, which is given off by the roots. Any other smell is irrelevant to the larva.
- The characteristic scent of every living creature is carried in molecules that leave the body in the form of sweat, breath, bodily wastes, and so on. With every step you take, you leave behind millions of sweat-borne molecules, which lay a scent trail.
- Butterflies, in the mating season, can attract a mate from 10 miles away by scent alone. The female butterfly carries a store of perfume, and she squirts minute fractions of it into the air.
- One of the oddest hearing systems in nature is that of the North American *ichneumon fly*. Its "ears" are in its feet, and it uses them to listen for the noise made by the larvae of the *horntail wasp* when they are chewing wood. The female fly locates the moth larvae with deadly accuracy, by running up and down tree trunks, listening with the hearing cells in her feet for the sound of chewing inside.
- Bees are sensitive to ultraviolet light, invisible to humans, which enables them to locate the position of the sun even when clouds obscure it. This helps them to navigate. Conversely, there are parts of the

Fascinating Creatures & Botanical Marvels

color spectrum to which bees are insensitive. For instance, they can see red only as black.

• Fireflies produce their light through a chemical process that takes place in an organ near the tip of the abdomen. Light producing chemicals and enzymes combine with oxygen to produce a bright glow. Fireflies, placed in lanterns, are used in some parts of the world for cheap lighting. You can read a book by the light of six fireflies.

• The *ant lion* uses its abdomen as a plow to dig a hole in the sand in its desert habitat. Then it sits in the hole and waits for its prey: ants and spiders. When one of these creatures approaches the hole, it disturbs the sand on the rim, and the ant goes into action. It places a grain of sand on its head and flicks it with unerring accuracy toward its target. This tumbles the creature into the hole, where it is quickly devoured.

• The *bombardier beetle* fires a jet of liquid, composed of hydrogen peroxide and quinoa at any would be attacker. This stream can cause severe pain to a man, if it gets into his eyes.

• The *golden tree frog* has a croak in winter that sounds like a mallet chipping rock, but in summer it sounds like a tinkling bell.

• A *white chamois* is considered in the Alpine region of Europe to be a portent of death. Superstitious hunters insist that any one who kills a white chamois will die within a year.

• The *ghost crab* is cream colored when it is on dry sand, but when the sand is wet the crab blends into a combination of gray, purple and brown.

• The *sulfur sponge* is the only animal life that can dissolve seashells into the original calcium from which they were formed.

• The *leaf bug* has legs and antennae the color and shape of leaves, has indentations on its body like vein marks on a leaf and hangs from branches, swaying in the breeze exactly like a leaf.

• The *Colorado River toad* has a croak that sound just like a ferryboat whistle.

• The katydid's world of hearing is much stranger than we can imagine, as it can hear sounds up to 45,000 vibrations per second. Compared to the human range of around 1700.

• *Phronima sedentaria*, a crustacean, lives in a glass house. It manufactures its transparent shelter out of a gelatin it obtains from salpa, a sea fish.

- The *warrior crabs* of Japan, look like the face of a samurai warrior, and are believed by natives to be the spirits of Japanese samurai warriors who died in battle.
- The seahorse of the Australian seas looks just like, and disguises itself as seaweed.
- The *calling hare*, which bleats like a sheep, is an expert ventriloquist. In fact it can make its voice appear to come simultaneously from several different points at once.
- A *silver butterfly* was once used in China as a currency.
- The *stepmother waterbug* deposits its eggs on the back of any passing bug. The eggs remain glued there until they hatch.
- The *tailless tenrec* of Madagascar produces the largest litter of any mammal - as many as 32 young at each birth.
- The *hawk moth*, which sucks nectar from blossoms, like a hummingbird, flies so swiftly that it can pollinate 200 different blooms in only 10 minutes.
- The *cabure* of South America is the smallest of all owls, yet for some strange reason, its hoot is a command that summons every single bird hearing the call.
- The sperm whale has an efficient depth gauge - the outer membrane of its eye - that flashes a warning to its brain when the whale has descended to the danger point.
- The *hercules moth* of New Guinea, measures 14 inches from wing tip to wing tip. It only lives for 14 days and takes no nourishment of any kind during its brief life span.
- The chipmunk can eat the deadly poisonous amanita mushroom, known as the death cup, without suffering ill effects.
- A tortoise is still alive in the garden at Nukualofa, in the Tonga Islands. It was a gift from Captain Cook - who left it there more than 200 years ago.
- The *pine slash beetle* has 7 or 8 mates, and builds separate nests for each of his families, for which he provides and protects.
- The *water bug frog* of Australia survives droughts of as long as 18 months by absorbing so much water during rainy season that it looks like an orange.
- The *leopard frog* leaps out of the water into the air and slays passing birds.
- The *snorkel snail* while under water breathes air through 2 tubes it extends to the surface.

Fascinating Creatures & Botanical Marvels

- The *water flea* has no problems with the female, as there are over 1 million females for each male.
- *Adelie penguins* drink only salt water 11 months a year. During their month of courtship they always change to fresh water - which they obtain by eating the Antarctic snow.
- *Backstroke bug* always swims on its back, and uses its hind legs as oars.
- The insect that fishes with a net. The *caddis fly* catches the small organism on which its feeds by spreading over the water of streams a net it spins out of silk.
- The *oecophylla ant* of West Africa creates its nest by sewing together leaves with silk threads.
- The *monarch butterflies* which breed in Canada and the northern U.S. migrate in the fall to the gulf states in formations so large they cover an area 430 miles in length and 40 miles in width. That's 17, 300 square miles of butterflies.
- The *purple sea snail* floats on the surface of water, supported by a raft of air bubbles, which the snail creates by rippling the sea with its foot.
- Caribou have hollow hooves that serve as suction cups on smooth ice.
- The African locust looks so much like a flower that the insects on which it feeds, settle on it in search of nectar.
- The *stalk eyed fly* of the tropics has its eyes at the tip ends of two long antennae.
- The larva of the *laternaria insect* frightens off predatory monkeys by camouflage that makes it look like a baby alligator - painted alongside its snout are rows of what looks like ferocious teeth.
- The *forest weevil* of Guiana, South America, when attacked by a fungus changes into a plant! The insect's shell (with fungus attached) remains transfixed to the plant. The weevil grows a new shell and then continues on its way.
- The Nile crocodile is the only animal in nature that makes an audible sound while it is still in its egg. The egg is buried in the ground, and the young crocodile's "honk" is a signal to its parent that it is time to dig up the egg.
- The *mimeciton* (a beetle) mimics the actions of ants so perfectly, that it is permitted to enter an anthill and get a working ant's share of the stored honey.

- The bird that plants a flower garden! The *gardener bowerbird* builds a garden enhanced by colorful blossoms and shells and a special hut next to the garden as a romantic setting in which to court its mate.
- The *giant spider crab*, which lives in the waters off the Ryukyu Islands off southwest Japan, stands three feet high and often weighs as much as forty pounds. His powerful legs can spread as wide as twelve feet, and his savage claws have killed a number of men.
- Tropical ants, when a flood sweeps down upon them, roll themselves into a huge living ball, which drifts upon the water - with their young safe and dry as its core.
- A *hermit crab* has a boarder and a servant. The boarder is an *anemone*, atop the crab's shell, and the servant is a *nereis*, a sea worm, that lives inside the crab's shell - and keeps it clean.
- The *berry butterflies* of Singapore in their caterpillar stage, group around the top of a stem to foil predatory birds by imitating the appearance of a poisonous berry.
- The *sponge crab* makes itself unappetizing to predators by cutting a piece of sponge that fits perfectly over its back.
- Predators do not harm the owl butterfly because its camouflage makes it look like an owl, which is a fierce bird of prey.
- The *African mason wasp* builds nests with many cubicles in which it imprisons spiders. Thus they have fresh meat whenever the need arises. These spiders evidently have an expiration date, because the wasps abandon each nest after one year, and build new ones.
- The *geometrid caterpillar* confuses its foes by impaling buds on its body spines so that it resembles a flower.
- The *oak leaf insect* looks amazingly like the scalloped leaf of the oak tree - and each autumn its color turns brown at the same time and rate as the leaves.
- The *mistletoe bird* of Australia feeds its young while hanging upside down.
- An aardvark will claw open a desert melon to drink the water inside the fruit. In doing so, it ingests a number of melon seeds. The seeds are excreted with the animal's dung. Since the aardvark buries its dung, the desert melon seeds find a place in the soil, and a supply of nourishing manure as well.
- *Toucans*, a family of tropical American birds, have extremely large bills. Often the bill is actually bigger than the bird's body.

Fascinating Creatures & Botanical Marvels

- The *Indian tailorbird* builds a most peculiar nest. The male of the species first finds two leaves close together near the end of a branch, and then using his bill as a needle, he sews them together with thin vegetable fibers.

- The *cicadas* (called seventeen year locusts, although they are not locusts at all) spends seventeen years eating and growing in underground passageways, then emerge for only two weeks of sunshine before dying. Seventeen years later, a new generation of cicadas will emerge from the ground.

- The bald eagle uses the same nest, and stays with one mate for a lifetime. Each year the eagle keeps this massive nest in repair by adding more material, so that, after a few years the nest will weigh more than a ton!

- . In the streams and lakes of South America is the most ferocious fish to be found anywhere in the world. The *piranha* is a small harmless looking fish with powerful jaws lined with razor sharp teeth. These fish always attack their prey in schools of several hundred. No animal of any size can withstand his or her vicious onslaught. A full-grown cow will be reduced to polished bones in less than two minutes.

The Strange Garden Of Nature

In the last chapter, we examined and pointed out the many strange and bizarre things connected with living creatures.

In this section, we are going to look at the many strange things in the garden of nature, where all manner of growing things exist and function in a wondrous world of beauty and mystery. Where unusual botanical marvels flourish, and oddities abound.

- The *Victorian water lily* grows in the Amazon region of South America. The English explorers who discovered and named the plant reported that it could safely support the weight of three men without being submerged. The round leaves of this jungle monster plant are over 8 feet in diameter.

- The *krubi flower* grows to a height of 15 feet in the jungles of Sumatra. The leaves of a well-grown specimen, when unfolded, can cover an area 45 feet in circumference.

- The *carrion flower*, which grows in Africa, is so named because it smells like decaying meat. This plant resembles carrion in appearance and odor, to fool carrion flies into laying their eggs in the petals. As the flies wander over the plant, their bodies pick up grains of pollen. The next time one of these flies is fooled by a carrion flower, these grains are inadvertently deposited on the second flower, thereby fertilizing the plant.

- The *cuckoopint*, an arum plant that grows in Europe, "tars-and-feathers" insects to assure a strange method of pollination. Insects enter its tube-like chamber and tumble down into a floral trap. As they attempt to leave, the slippery walls of the chamber, in which they are trapped all night, "tar" them. In the morning, the stamen further up in the tube trembles and "feathers" the insect with pollen. The flower then opens up and allows the pollen-laden insect out.

- There are many species of pitcher plants, but they all have hollow leaves that act as a death trap for invading insects. The inside walls of the slender chamber are covered with thin hairs. The sharp, downward pointing bristles prevent the struggling insect from moving in any direction but down into a pool containing a narcotic drug which incapacitates the doomed creature.

- Flowers do not appear on the giant *saguaro cactus* until the plant is 60 or 70 years old. The largest specimens are over two hundred years old and weigh over 10 tons.

- The *"sensitive plant"* seems to be capable of experiencing both fright and fatigue. When any of the leaves are touched, the plant seems to wilt from fright before your very eyes! After a short time in the wilted condition, the plant will return to normal. If the plant is touched repeatedly in a short time, it acts as if it had become exhausted.

- The *partridgeberry* is always a botanical siamese twin of nature. Each berry develops from two flowers.

- The *samanea saman* is a tropical tree that rains. The tree collects moisture in its pods during the day and each evening dumps it in the form of a heavy rain.

- The *barometer bloom* of South America always sends forth a few blossoms on the day before a rainstorm.

- The *telegraph plant* of Asia has leaves that flutter constantly even when there is no breeze.

- The *giant puffball* produces 7,000,000,000,000 spores, fortunately only one becomes a puffball, and the others all die.

- The *goats beard flower* produces many flower clusters, each of which contains more than 10,000 blossoms.

- *The candlesticks of the sun.* A plant in Central Australia that grows a candle shaped flower only once in every seven years.
- *The woman's tongue* of Zanzibar is a plant with pods full of seeds that rattle continuously.
- The *entada pursoetha* (a giant bean-pod) grows to a height of 4 or 5 feet and is so sturdy that the people of Eastern Pakistan use it as a stairway to their dwellings.
- *The living torches (puya raimondi)* grows in the cordillera mountains of Peru and has a shaft of blossoms 35 feet high that is so saturated with resin that shepherds light them to illuminate the countryside.
- Berries of the yew tree and the branches themselves are highly poisonous to cattle - yet birds eat them without ill effects.
- *Fire coral* of the Red Sea is intensely hot. Touching it causes painful burns, blisters, and permanent scars.
- *The compass plant* of South Africa is used as a guide by travelers because it always leans toward the north.
- The calopogon orchid dumps to the ground any insect that alights on it, and then spills its pollen on the insects back.
- *The rafflesia flower*, which grows in the Sumatran rain forest, is a parasite flower that grows on the exposed root of the cissus vine. After a rafflesia dies, the flower decays into a sticky liquid, in which the seeds float. When an elephant walks through the gluey substance, it adheres to his feet, which annoys the animal. The cissus root is the perfect tool for the elephant to use to wipe the stuff off its feet, and he thereby leaves a deposit of rafflesia seeds on the vine, which soon sprouts into new rafflesia plants.
- The members of one species of sequoia known as big trees sometime grow as high as 325 feet, with trunks up to 30 feet thick. Some of these trees are believed to be more than 4,000 years old!
- The *squirting cucumber* is a Mediterranean plant that literally explodes to disperse its seeds. The cucumber swells with liquid to the bursting point, then explodes and propels the seeds as far as 40 feet from the plant.
- The *parasitic mistletoe* sucks the life from its host tree, and thus brings about the death of both. As birds feed on the plant's waxy white berries, the seeds stick to their bills and feet. The birds scrape off the seeds on the bark of trees, and when the seeds germinate, the roots bore into the tissue of the newfound host.

- The seeds of the *Indian lotus* enjoy a short river cruise before they settle down. The plant takes root in the river bottom, with its stalk extending upwards through the water. When the seeds are ripe, a "receptacle" breaks off from the stock and the "ark" begins floating downstream. Eventually, the sprouted fruits are dislodged from the pod and sink to the bottom of the river, where the cycle begins anew.

- The *Indian fig tree* (or *banyan* tree) grows downward from a rooting place high above the ground. Birds into the branches of a host tree drop the seeds of the banyan tree, where they germinate. Rope-like shoots descend from the sprouting seed and take root in the soil at the foot of the tree. These shoots thicken until they form trunks. The banyan tree may develop as many as 320 trunks and cover a ground area of 2,000 square feet.

- The *molocca bamboo* can grow two feet in just 24 hours! Some species grow in tufts to a height of 100 feet. The blades provide paper, fuel, furniture, utensils, plumbing pipes, ship masts, and construction material for homes.

- The *elephant tree*, a small tree in northern Mexico, discharges a cloud of fetid-smelling oil to protect its leaves from hungry herbivores! A mist of oil inundates any animal that tries to nibble on one of these fruits or leaves, as foul as a skunk's spray. The mist actually consists of tiny particles shot in a jet from openings in the bark.

- There is a family of plants that has developed a highly effective form of camouflage. These plants resemble rocks so closely that they are called *"living rocks."* These rock mimics are found in the barren, rock-strewn wastes of southern Africa.

- The jumping bean is the seed of the *yerba de flecha*. A Mexican shrub related to the rubber tree. A certain species of moth deposits eggs in the blossoms. When the egg hatches, the caterpillar bores into a seed. The movement of the bean results from the larva's attempts to escape direct sunlight, which would render the inside of the seed too hot for comfort. The larva shifts about inside the seed, turning and tumbling the seed until it finds a patch of shady ground.

- The *aspidistra* produces small, bell shaped flowers at ground level. As a snail stops to nibble on the petals, it picks up grains of pollen, which are subsequently transferred to another aspidistra plant.

- The *passionflower* was so named because the structure of the plant was thought to be symbolic of the crucifixion of Christ. The flower has 10 petals, five stamens, three styles, a fringed corona, and coiling tentacles. Symbols for, respectively, for the 10 faithful apostles, the five wounds of Christ, the three nails, the crown of thorns, and the scourges.

- Skunk cabbage each spring pushes up through the frozen ground rolled up like a cigar. The temperature inside the plant is 25 degrees warmer than the outside air.
- The *weather thistle* forecasts rain by closing its golden blossoms and announce the coming of fair weather by opening them.
- The *araucara* trees of Chile, which furnish wood famed for its hardness, are never felled for lumber until they are 500 years old.
- The fruit of the bead tree of India feeds sheep, goats, and birds - yet it is highly poisonous to man.
- The blossoms of the fig tree are never visible. The blossoms are inside the fig, where they develop into seeds.
- The plant that is all seed! *Bryophyllum* re-creates itself when any fragment of its leaf or steam falls to the ground.
- The *sweet-after-death plant* has no fragrance while it is alive, but after its leaves have died and withered they have the sweet smell of vanilla.
- The lush green leaves of the *gabor bush* of the Sahara Desert fatally poison any living creature that eats them.
- Allspice gets its name from the fact that its berry combines the mingled fragrance of cloves, nutmeg, and cinnamon.
- The sassafras tree bears 3 different shaped leaves on the same branch.
- Melons growing in the Sahara desert are so bitter they are fed to thieves for punishment - yet camels and gazelles consider them a delicacy.
- *Vitex Agnus Castus*, the branches of which are used to make baskets, produces a flower that was eaten in ancient Greece by young girls in the belief it would enhance their virtues.
- Trees growing along the Columbia River gorge in Oregon are buffeted by such strong winds - invariably from one direction - that all their branches extend leeward.
- The *raspberry Jam tree* of Australia is so called because its wood smells just like raspberry jam.
- The *duck orchid* of Australia, swaying on its slender stem, actually looks like a duck in flight.
- The *Star of David orchid* consists of two triangles inverted one atop the other - resembling the Star of David.

- The *chocolate orchid*, which grows in Europe, has both the color and scent of chocolate.
- The *agave* plant of Brazil has its only leaves on the ground, but its stem grows to a height of 30 feet. It dies as soon as it reaches its full height.
- The *hedgehog cactus* of Mexico is pleated like an accordion, so it can expand with water in preparation for periods of drought.
- The *queen-of-the night cactus* that only blooms at midnight has a perfume so powerful that it can be detected half a mile away.
- *Witch Hazel* was so named because early Americans believed that twigs from the plant had the power to locate underground springs and minerals when used as a divining rod.
- The *cornel shrub* grows its blossoms before it sprouts leaves.
- *Sweet William*, a flower, and *stinking Willie*, a weed, both were named for the same man William, Duke of Cumberland, who defeated the Scots in the Battle of Culloden. The grateful English named a flower in his honor, and the Scots gave his name to a weed!
- The *curtain fig tree* of Australia is a giant tree with branches so dense they form a natural curtain.
- The *resurrection plant* moves in search of water. A desert growth found in arid regions of America and the Near East, owes its name to its extraordinary ability to come to life again from a seemingly dead and shriveled state. When moisture is scarce, the plant pulls up its roots and withers into a dry, ball like mass. This withered mass is carried along the ground by the wind, and can remain in a dormant state for years, if no water is found. However, when water is located, the plant sinks roots into the wet ground and springs to life again.

6

The Paranormal & the Supernatural

The Near Death Experience

The near death experience (NDE) is a term used to describe a mystical experience that happens to people who have been declared clinically dead, only to return to life. An extensive poll conducted by the George Gallup organization in 1982, uncovered an estimated eight million near-death experiences, which some say glimpses into the next world.

What is it like to die, and then be revived? While each person has a different and astonishing story to tell, their experiences contain striking similarities, and all agree that to come back from the dead is a life changing experience.

Dr. Raymond Moody Jr. M.D., the author of *LIFE AFTER LIFE*, has studied more than one hundred subjects who have experienced "clinical death" and been revived. Their accounts of this experience are startlingly similar in detail.

A man is dying and, as he reaches the point of greatest physical distress, he hears himself pronounced dead by his doctor. He begins to hear an uncomfortable noise. A loud ringing or buzzing, and at the same time feels himself moving very rapidly through a long dark tunnel. After this, he finds himself outside his own physical body. Soon, other things begin to happen. Others come to meet and help him. He glimpses the spirits of relatives and friends who have already died, and a loving, warm spirit of a kind he has never encountered before - a being of light - appears before him.

Dr. Moody has examined actual case histories that reveal there is indeed life after death. People declared clinically dead give descriptions so similar, so vivid, and so overwhelmingly positive that they may change mankind's view of life, death and spiritual survival forever!

It is research such as Dr. Moody presents that confirm what we have been taught for two thousand years - that there is life after death. A dying patient continues to have a conscious awareness of his environment after being pronounced clinically dead.

Dr. Elisabeth Kubler-Ross, M.D., writes: "This very much coincides with my own research, which has used accounts of patients who have died and made a comeback, totally against our expectations and often to the surprise of some highly sophisticated, well-known and certainly accomplished physicians."

All of these patients have experienced a floating out of their physical bodies, associated with a great sense of peace and wholeness. Most were aware of another person who helped them in their transition to another plane of existence. Loved ones who had died before them greeted most.

Dr. Kubler-Ross, points out that "we have reached an era of transition in our society. We have to have the courage to open new doors and admit that our present-day scientific tools are inadequate for many of these new investigations."

P.M.H. Atwater, author of *BEYOND THE LIGHT,* gives startling new evidence of life after death - from visions of Heaven to glimpses of Hell. In his well-researched book he gives us an overall view. Regardless of descriptive variants, the overall pattern remains the same:

- *"A sensation of floating out of one's body, often followed by an out-of-body experience where all that goes on around the vacated body is both seen and accurately heard.*
- *Passing through a dark tunnel or black hole or encountering some kind of darkness. This is usually accompanied by a feeling or sensation of movement or acceleration. Wind may be heard or felt, or a swooshing sound may predominate.*
- *Headed toward and entering into a light at the end of the darkness, a loving light of warmth and brilliance, with the possibility of seeing people, animals, plants, lush outdoors and even cities within the light.*
- *Greeted by friendly voices, people, or beings, who may be strangers, loved ones, or perhaps religious figures. Conversation can ensue; information or a message may be given, as part of the scenario.*
  Seeing a panoramic review of the life just lived, from birth to death or in reverse order, sometimes becoming a reliving rather than a dispassionate viewing. The person's life can be reviewed in its entirety or in segments. This is often accompanied by a feeling or need to assess gains or losses made during the life, so the individual

can be aware of what was learned or not learned. Other beings can participate in this assessment or offer advice. It is possible for such memories to be open-ended and to include all existent knowledge, not just personal revelations.
- *A different sense of time and space, discovering that time and space do not exist, along with losing the need to recognize such measurements as either valid or necessary.*
- *A reluctance to return to the earth-plane, but invariably coming to realize that either one's job on earth is not finished or a mission is yet to be performed before one can return to stay.*
- *Disappointment at being revived, feeling a need to shrink or somehow squeeze to fit back into the physical body. There can be unpleasantness, even anger or tears, at the realization that one is now back in his or hers body and no longer on the other side. Fear of death either subsides or disappears altogether."*

Dr. Melvin Morse, M.D., in his excellent book *Closer to the Light*, reveals two marvelous discoveries.

It has been proven that a person actually needs to be near death to have a near death experience. This finding silenced many skeptics who had said that these events were just hallucinations that any seriously ill patient could have. By scientifically comparing the experiences of seriously ill patients with those who had been on the brink of death, Dr. Morse, was able to determine that one does need to cross that threshold before glimpsing the other side.

Also he has been able to isolate the area in the brain where near death experiences occur. This area, close to the right temporal lobe, is genetically coded for the NDE. Dr. Morse and his researchers explored whether this could be the "seat of the soul," the area that holds the vital essence that makes us what we are.

In some cases, those who are having the NDE, while out of their body, float outdoors, or even travel to see loved ones. Some of these "invisible" experiencers were at home when the phone call came about their death, and later, they were able to give exact accounts of - who was there, what they were wearing, what was said - and to do so without flaw.

The average near death survivor comes to regard him or herself as an immortal soul currently resident within material form so lessons can be learned while sojourning in the earth-plane. They now know they are not their body; they are a living soul, a child of God.

The world is the same, but not the individual. This also includes children. As a general rule, though, most near death survivors do not recognize the extent to which they have changed. Often, the most revealing stories come from the families. Some experiencers often go on learning binges afterward, craving knowledge. Some enroll in schools and self-development classes, and take as many field trips as their budgets allow for hands on learning. They gravitate to teaching and counseling roles, or to the areas of religion, spirituality, and healing. P.M.H. Atwater, in his research, has discovered a number of typical physiological aftereffects that affect about 90 percent of near death survivors, which include: They tend to look and act younger, have brighter skin, and eyes that sparkle. Have more energy. Become more sensitive to sunlight. Are more sensitive to sound and noise levels. Are more open and accepting toward the new and the different. Regard things as new even when they're not. Handle stress more easily and heal faster, and exhibit changes in brain functioning.

Many of these people also experience physiologically, or even metabolic changes which include a greater sensitivity to household chemicals, A new preference for open doors, windows, and shades, the ability to hear words and music in the air when no one is nearby. They seem to attract animals, birds, and children just by their presence. Plants seem to grow better around them. Latent talents tend to surface. They all seem to have heightened sensations of taste-touch-texture-odors, sensitivity to electricity and geomagnetic fields. Sparkles or balls of energy are sometimes seen in the air, increased sensitivity to temperature, pressure, air movement, and humidity. Their body energy interferes with electronic equipment, light sources, security systems, and the like. Extrasensory perception and other psychic abilities become routine. Experiencers often develop healing hands, and exhibit a charismatic aura around them. Some gain the ability to know the future. They become more creative. The body seems to assimilate food more quickly. And last, but not least, experiencers become more orgasmic.

Death loses all meaning and relevance to those who survive the NDE. The individual now knows death does not end anything, and is only a change in awareness. The individual becomes so magnetic that people and animals are drawn to him or her. The individual seems divinely protected and guided.

The typical near death survivor was dead for ten to fifteen minutes. It is not uncommon, however, to hear of clinically dead experiencers who revived thirty minutes to an hour, or even several hours later!

One woman was dead for twelve hours, and another for sixteen. Both "woke up" in a morgue. This is amazing to me since the medical community cautions that without sufficient oxygen, the brain can be permanently damaged in three to five minutes.

The near death phenomenon seems to stimulate the brain hemisphere that was not previously dominant. There is also, according to neuroscientist Arnold Scheibel of UCLA, an observable movement in the brain, structurally, toward data clustering and creative invention - as if the experiencer were developing a more synergistic type of neural network - thus advancing the potential of whole-brained behavior.

The philosopher Plato's writings are full of descriptions of death. For instance, Plato defines death as the separation of the incorporeal part of a living person, the soul, from the physical part, the body. What is more, this incorporeal part of man is subject to many fewer limitations than is the physical part. Hence, Plato specifically points out that time is not an element of the realms beyond the physical, sensible world. The other realms are eternal, and in Plato's striking phrase, what we call time is but the "moving, unreal reflection of eternity."

Plato discusses in various passages how the soul, which has been separated from its body, may meet and converse with the departed spirits of others and is a guided transition from physical life to the next realm by guardian angels.

Plato remarks that the soul that has been separated from the body upon death can think and reason even more clearly than before, and that it can recognize things in their true nature far more readily. Furthermore, soon after death it faces a "judgment" in which a divine being displays before the soul all the things - both good and bad - which it has done in its life and makes the soul face them.

The Tibetan Book of the Dead is a remarkable work compiled from the teachings of sages over many centuries. The book contains a lengthy description of the various stages through which the soul goes after physical death.

In the Tibetan account, the mind or soul of the dying person departs from the body. He may hear alarming noises described as roaring, or whistling like the wind, and finds himself surrounded by a misty illumination.

He is surprised to find himself out of his physical body. He sees and hears his relatives and friends mourning over his body and preparing it for the funeral and yet when he tries to respond to them they neither hear nor see him. He does not yet realize that he is dead, and he is confused. He asks himself whether he is dead or not, and when he finally realizes that he is, wonders where he should go or what he should do. A great regret comes over him, and he is depressed about his state. For a while he remains near the places with which he has been familiar while in physical life.

He notices that he is still in a body - called the "shining body" - that does not appear to consist of material substance. Thus, he can go through rocks, walls, and even mountains without encountering any resistance. Travel is almost instantaneous. Wherever he wishes to be, he arrives there in only a moment. If he has been in physical life blind or deaf or crippled, he is surprised to find that in his "shining body" all his senses, as well as all the powers of his physical body, have been restored and intensified. He may encounter other beings in the same kind of body, and may meet what is called a clear or pure light. All deeds both good and bad are reflected for both him and the beings judging him to see vividly.

It is quite obvious that there is a striking similarity between the account in this ancient manuscript and the events described as NDE.

What we now know as near death experiences have been reported since the beginning of recorded history. In the New Testament (2 Cor. 12:1-4) Paul describes one that he had. Pope Gregory the Great in the sixth century collected these experiences as proof of life beyond.

Carol Zaleski, a prominent Harvard theologian, finds near death experiences in Greek, Roman, Egyptian, and Near Eastern religions. In her book *OTHER WORLD JOURNEYS,* she states; "that some cultures see death as a journey whose final goal is the recovery of one's true nature,"

Visions Of Heaven & Glimpses Of Hell

A Study done by Nancy Evens Bush contains stories that link the Light with rebirth. Bush based her reports on accounts in the archives of The International Association for Near Death Studies (IANDS) that was founded in 1977 by Dr. Raymond Moody, Dr. Bruce Greyson, Dr. Michael Saborn, and Dr. Kenneth Ring. The purpose of the association is to bring an interdisciplinary approach to the near death research. It now has hundreds of chapters worldwide.

In one of these accounts, a four year old girl, using a flashlight to go down the cellar stairs, stepped off the edge of the wrong side of the landing and fell to the cement floor far below. At a later age, she described what happened next:

> *"The next thing I was aware of was being up near the ceiling over the foot of the stairs. The light was dim at first. I saw nothing unusual. Then I saw myself lying, face down, on the cement, over to the side of the stairway. I was a little surprised, but not at all upset at seeing myself that way. I watched and saw that I didn't move at all. After a while, I said to myself, 'I guess I'm dead.' But I feel good! Better than I ever had. I realized I probably wouldn't be going back to my mother, but I wasn't afraid at all.*
>
> *I noticed the dim light growing slowly brighter. The source of light was not in the basement, but far behind and slightly above me. I looked over my shoulder into the most beautiful light imaginable. It seemed to be at the end of a long tunnel which was gradually getting brighter and brighter as more and more of the light entered it. It was yellow-white and brilliant, but not painful to look at even directly.*
>
> *As I turned to face the Light with my full 'body,' I felt happier than I ever had before or have since. Then the Light was gone. I felt groggy and sick, with a terrible headache. I only wanted my mother, and to stop my head from hurting.*

To understand the Light further, it is necessary to comprehend the power it can have to illuminate our lives. The experience of near death researcher Michelle Sorenson as recorded in the book *Closer To The Light* by Melvin Morse, M.D., and Paul Perry, illustrates this point:

> "Suddenly I was above my body, looking down from a corner of the room. I felt a wonderful warmth with no chills. A man was standing behind me. The warmth seemed to come from that person and spread around me. I did not turn around. I stared in relief at my form on the bed. I was at peace. I knew I was dead.
>
> Then I thought 'I should have done this sooner!' Over the years, I have had a hard time explaining how this man talked to me. Yet he did, and the communication was so warm and loving and so peaceful, that I knew the radiant white light was his love. He knew what I had been through, and his compassion put me at rest.
>
> Looking down at myself on the bed below, I saw my friend put her hand to my body's forehead and then neck to find a pulse. She was screaming. Other people were shouting. 'She's dead, she's dead.'
>
> I saw my mother's face and my brother's face. He was over-seas and they were calling him. I saw a whole network of phone lines, with people's faces on the phones. I felt sad that they were upset. But felt that they would get over it. Even my mother and father would want me to have release from the pain I was having.
>
> '"But look what you are missing," the voice said.
>
> '"I saw a tall blond man walking with two children. The little girl jumped up and down and her curls shook. The other was a boy. I recognized this as being my future family. I felt a longing for my husband and children even before I had met them!
>
> The bliss I felt as a dead person suddenly felt temporary. I began to waiver about the joys of being dead before I had even experienced the fullness of life. 'Yes I want to go back,' I said. And I went back."

The Light changed Michelle's life, "I realized that death was not to be feared," said Michelle.

By the way, Michelle is now married to a former basketball player who is blond. They have two children, a boy and a girl.

There have been instances in which the Light has intervened to save children.

In 1986, in Cokeville, Wyoming, David Young held 156 children hostage, and he had a bomb. He said he was going to kill them all. Young blew up the entire school - yet none of the children were hurt.

Later, the children described seeing people of light who directed them to safety before the explosion occurred. These beings of light told the children where to go to avoid being killed!

One little girl said, "The people of light were standing there above us.

Another said. "They were dressed in white, and were bright around the face."

One boy said, "the angel told me to take my little sisters over by the window, and keep them there."

Could this light, seen by the person having a NDE, be a physical manifestation of our guardian angel? Many children have described guardian angels, which are blond or dressed in white, as people of light who have come to escort them to heaven, or to save their life.

For example, one child having a NDE, left her body, went up a long tunnel, saw a paradise of light, was engulfed by a "Godlike light", and met her guardian angel named Sarah.

According to science, death should be the end of life and light. With the extinction of consciousness, there should be an absence of light. This "Light of love" is the core of the near death experience and cannot be explained by any scientific theories.

In Paul's experience in (Acts 9:3-6) the light was so bright that he was blinded for three days. This appearance of light was only one of fifty-two instances where light is mentioned in the Bible, the "true light" identified as Christ in John 1:9.

Dr. Maurice S. Rawlings, M.D., in his book *Beyond Death's Door*, gives a very interesting account of a man that goes all the way to heaven, and then returns.

A retired accountant had an unexpected meeting with this Light of love in our next narration:

> *"The next thing I remember, I was looking down from the ceiling in the intensive care unit. There was this young doctor in a white coat and two nurses bent over me. The doctor was yelling.*

Then I was going through this dark passage. I didn't hit any of the walls and at the other end I walked out into an open field. On the far side was this endless white wall that had three steps leading up to a doorway. I entered. Up on the landing sat the dazzling figure of a man in glowing white robe. He seemed to be looking down into a big book and studying.

As I approached I felt an overwhelming reverence and I asked, "Are you Jesus?"

He said, "No, you will find Jesus and your loved ones beyond that door." Then he looked again in his book and nodded. "You may go on through."

On the other side of the door I was amazed to find a brilliantly lit city.

Then I saw two figures walking toward me and I knew immediately it was my father and mother.

Then I noticed that one building was much larger than all of the others. It looked like a football stadium with the open end radiating a brilliant light.

"In there is God," they said. I will never forget that scene.

As we approached the place were Jesus was I felt as if I were hit and everything became dark. A jolting power went through my body and it hit me in the chest again, arching my body upward. I opened my eyes and I was back in my body."

This was yet another positive identification of Jesus or his angels as the beings of light. Which is of tremendous importance to these people, reinforcing the very core and conviction of their faith.

Dr. Kenneth Ring, in his research, discovered that there were variations of the NDE, depending on how people nearly died. Suicides tended to have truncated experiences that ended before they reached or even noticed the light at the end of the tunnel. While in the darkness, though, most were able to understand that suicide was wrong. They were left with a message that life did have a purpose and that they were going to be sent back because they had something to do.

If a person had been drugged or delirious when they were near death, they had no NDE at all. (That they could remember.)

Not all near death experiences are good or positive, or have a loving light waiting for them at the end of the tunnel. A significant number of people have a Hellish NDE. These were trips to lands of fiery lakes and demons! There was no happy ending in a sublime light. Returning people would say they'd found themselves suspended in emptiness with nothing to see or hear or contemplate but the horrible suspicion that life, for them, ultimately went nowhere.

Many of these people had transformations in there life styles following such an experience. One man said, "I don't ever want to go to that place again!" And he started going to church.

Another said, "I thank God that he gave me another chance - before it was to late!"

Hell itself has always been the believer's worst nightmare, always wondering if it is safe to die, continually reaching for reassurance.

Maurice S. Rawlings, M.D., author of *To Hell and Back*, has researched this area of the near death experience thoroughly and in depth. He speaks with an unusual degree of perception and honesty about his own patients who have died and gone to Hell, before returning to their own bodies.

For example, George Godkin of Alberta, Canada related a despairing near death experience:

> *"I was guided to the place in the spirit world called Hell. This is a place of punishment for all those who reject Jesus Christ. I not only saw Hell, but felt the torment that all who go there will experience.*
>
> *The darkness of Hell is so intense that it seems to have a pressure per square inch. It is an extremely black, dismal, desolate, heavy, pressurized type of darkness. It gives the individual a crushing, despondent feeling of loneliness.*
>
> *The heat is a dry, dehydrating type. Your eyeballs are so dry they feel like red-hot coals in their sockets. Your tongue and lips are parched and cracked with the intense heat. The breath from your nostrils as well as the air you breathe feels like a blast from a furnace. The exterior of your body feels as though it were encased within a white-hot stove. The interior of your body has a sensation of scorching hot air being forced through it.*
>
> *The agony and loneliness of Hell cannot be expressed clearly enough for proper understanding to the human soul; It has to be experienced."*

Most of the negative experiences do not contain this much graphic information. And very few ever actually reach the depths of Hell as this man did. There are some surprising similarities, and many variations that we shall explore.

In the "good" NDE, the person goes up, or through a dark tunnel. In the "bad" NDE, the person falls down into a spinning vortex, or is pulled into a dark hole.

P.M.H. Atwater gives a good example of this type of experience in the words of Gloria Hipple:

> *I recall being pulled down into a spinning vortex. At first, I did not know what was happening. Then I realized my body was being drawn downward, headfirst. I panicked and fought, trying to grab at the sides of the vortex. I tried to see something, but all there was to see was this cyclonic void that tapered into a funnel.*
>
> *I kept grabbing at the sides but my fingers had nothing to grasp. Terror set in, true terror. I saw a black spot, darker than the funnel and like a black curtain, falling in front of me. Then there was a white dot, like a bright light at the end of the funnel. But as I grew closer, it was a small white skull. It became larger, grinning at me with bare sockets and gaping mouth, and traveling straight toward me like a baseball. Not only was I terrified; I was really livid, too. I struggled to grab hold of anything to keep me from falling, but the skull loomed larger. "My kids, my baby is so little. My little boy, he's only two years old. No! My babies need me! No! No! No!"*
>
> *The skull shattered into fragments and I slowed in movement. A white light was in place of the skull. The black spot was gone. I felt absolute peace of mind and sensed myself floating upward, and I was back."*

Her Hellish experience transformed her life. She is no longer dependent on outer circumstances and material possessions. She now speaks effusively of God and angels, and seeks the realization of greater truths. She has found inner peace.

In another case variation, Sandra Brock was assaulted by entities awaiting her arrival in the tunnel.

Bruce Greyson, and Nancy Evans Bush, president of IANDS, has completed a descriptive study of fifty terrifying near death cases over the past few years. Which include lifeless or threatening apparitions, barren or ugly expanses, threats, screams, silence, danger, violence and torture, a feeling of cold, or of temperature extremes, and a sense of Hell.

In separate clinical deaths occurring in the same person, it's always bad-to-good conversion. No cases on file of good-to-bad. It appears there's nothing like a little bit of hell to dramatically change life's purpose and attitude.

Some of the phrases used in descriptions of Hell-like experiences are as follows:

- I saw twisted faces grimacing as they stared down at me.
- Obviously, I was in Hell itself.
- A black silhouette of a demon, had come to get me.
- As soon as I had left my body, I saw two demons waiting.
- Devils are ready to drag my soul down to Hell.
- As I was sucked into the black vortex, hands grabbed me.

Everybody has fears of things unseen, fears of things new and untried. Doubts are there. But believers who have experienced clinical death are no longer afraid. They have already glimpsed the glory. Paul, who had a personal experience with what might have been clinical death when he was stoned at Lystra, tells us that we should not be afraid:

"Now we look forward with confidence to our heavenly bodies, realizing that every moment we spend in these earthly bodies is time spent away from our eternal home in heaven with Jesus." (2 Cor. 5:6-9 TLB) Our earthly bodies, the ones we have now that can die, must be transformed into heavenly bodies that cannot perish but will live forever. When this happens, then at last this Scripture will come true - Death is swallowed up in victory. O Death, where then your victory? Where then your sting? For sin - the sting that causes death - will all be gone: and the law, which reveals our sins, will no longer be our judge. How we thank God for all of this! It is he who makes us victorious through Jesus Christ our Lord."
(1 Cor. 15:53-57 TLB)

Angelic Encounters

Samuel Frame of Staunton, Virginia, and his neighbors were accustomed to taking their grain to a mill operated by a man named Palmer. They had stored their grain at the mill as usual in 1870, waiting for the fall rains to raise the river so that Palmer could grind their wheat into flour.

But on the night after Frame took his grain to the mill for storage, he had a vivid dream, in which "an angel appeared."

He told his family the next morning that he had dreamed of a "lady in a bright gown" who warned him to remove his grain soon; for a flood would destroy the mill. Frame hurried back to the mill the next day and removed his grain. When he urged his friends to heed the warning and remove their grain, too, they laughed at him. That night there was a veritable cloudburst that swept away the mill and everything in it, exactly as the "bright lady" in the dream had predicted. On Samuel Frame's tombstone there is this legend:

"Samuel Frame, buried on this farm. He was warned by an angel in a dream, September 22, 1870, to remove his wheat from Palmer's Mill, now Spring Hill, which he did the following day."

According to The Bible there is an intelligent order of created beings that serve God, called angels. They have to do with God, and they also have to do with man. Angels are in many ways very much like men. They possess attributes of intellect and personality and will similar to those of man. Angels can speak. So like men are Angels that it is possible to sit and talk with one and not to know that he is an angel, but to take him for a man. Hebrews 13:2 reads, "be not forgetful to entertain strangers, for thereby some have entertained angels unawares."

George Washington spoke of his guardian angel often, and many times credited his success at Valley Forge to an inspiring visit from an angel.

Abraham Lincoln often called upon the healing powers and wisdom of angels to help guide him and unite the nation during a time of bloodshed and devastation.

Who are these cosmic creatures? How many are among us now?

Belief in Angels is found in the history of all nations. The ancient Egyptians, Phoenicians, Greeks, and others all expressed their belief in angels. They believe that two angels are assigned to each person. The angel on the right hand records all your good deeds. The angel on your left records all your evil deeds. All the major religions of the world - Christianity, Judaism, and Islam - accept angels as an intervening force between a Supreme Being and the human race. Angels also are major figures in Buddhism and Hinduism.

The Hebrews taught there were four great angels: Gabriel, who reveals the secrets of God to man; Michael, who fights for and avenges God's foes; Raphael, who receives the departing spirits of the dead; and Uriel, who will summon everybody to the judgment.

The Paranormal & the Supernatural

The earliest archaeological evidence of angels to date appear on the Stela of Ur-Nammus (2250 B.C.) and shows angels flying over the head of this King while he is in prayer.

The Old Testament contains over 300 references to angels and their duties here on Earth.

The earliest appearance of angels is found in Genesis, when an angel armed with a flaming sword drove Adam and Eve out of Eden and was instructed to guard the gates so they could never return.

In the Psalms we read that He "will command his angels concerning you to guard you in all your ways."

In the book *There's an Angel On Your Shoulder,* Kelsey Tyler gives a very interesting example of an angelic encounter in everyday life. An encounter that Chris Smith and his family will never forget.

Chris Smith had driven the two-lane highway etched into the side of a mountain many times. Sheer drops of several hundred feet bordered the road.

The pick-up truck was loaded with lumber, and Chris felt the lumber began to shift. He slowed the vehicle down enough to prevent the load from spilling over. A lumber spill on this road could cause a serious accident. He began to pray that the load would stay on the truck.

As Chris looked into his rear-view mirror, he could see several impatient drivers had come up behind him. But with the lumber starting to shift, he had to keep his speed down. There were only inches separating the road from the canyon's edge, and no place for Chris to pull over, and let the cars behind him pass. He glanced once more in his rear view mirror, and took his eyes off the road for just a second. When he looked again, his truck was heading off the roadway. Chris decided not to slam on his brakes. If the car behind him hit him, the impact could send him over the cliff.

The Earth under his front right tire gave way, and the pick up-truck began tumbling down the mountainside into the canyon.

As his vehicle continued crashing and rolling down the side of the mountain, Chris felt he would die, but his biggest fear was for his daughter Kailey, who was in the back seat.

The truck came to a rest some 500 feet down the mountain. He called out to his daughter, "Kailey, where are you?" But he could hear only the sound of the wind. Kailey's car seat was still strapped to the back seat; the body harness was still snapped in place. But Kailey had disappeared.

Chris, his body almost paralyzed with pain, knew that if Kailey had been thrown from the truck during the fall, she would almost certainly have been killed.

Chris began to pray. "Please help me find my little girl, oh God don't let her be dead."

Moving as fast as he could, he made his way up the mountainside. He called her name out every few feet. But there was no answer.

When he was some 30 feet from the top, he heard her calling his name.

As Chris reached his daughter, she was sitting cross-legged on top of a soft bush.

At about this time a medical helicopter landed on the road about 25 feet away. Chris was placed in intensive care and given a slim chance to survive.

Kailey, who had been taken to pediatric unit for observation, told her mother, Michele, what had happened.

"We were driving and then we started to fall, then the angels took me out and set me down on the bushes." She said.

"Tell me more about the angels," Michele urged her daughter.

"They were nice. They took me out and set me on a soft bush." Kailey said again.

Over the next few days Chris made a miraculous recovery, which the doctors said was nothing less than a miracle.

We note that in the Old Testament, angels appear often. They are cast as working behind the scenes in the drama of world events as agents of God to promote His program.

Angels, who were like men, came to Abraham - the patriarch of both Jewish and Islamic religions - and to members of his family. Three of these angels came to Abraham and his wife, Sarah, when the couple was more than 90 years old to tell them that they were going to have a child.

In Chapter 19 of Genesis, there is an account of a visit by three angels to Abraham's nephew Lot: "The angels came to Sodom in the evening, and Lot was sitting in the gate of Sodom. When Lot saw them, he rose to meet them, and bowed himself with his face to the earth, and said, 'My lords, turn aside, I pray you, to come to your servants' house and spend the night and wash your feet; then you may rise up early and go your way. So they turned aside to him and entered his house, and he made them a feast and baked unleavened bread and they ate."

These were not men; they were angels, yet they had feet and they needed to wash them after their walk, and Lot prepared a feast for them, "and they ate,"

The angels told Lot that the Lord had sent them to destroy five cities in the region because "the outrage of Sodom and Gomorrah is so great, and their sin so grave."

Lot had just finished feeding the two angels who "had not laid down yet." This verse seems to indicate that angel's lie down and rest - or even sleep.

"The men of Sodom," gathered outside of Lot's house. And shouted "bring them out so that we may be intimate with them."

But Lot said, "do not commit such a wrong act." But the crowd moved forward to "break down the door." The angels stretched out their hands and pulled Lot into the house, "and the crowd who were at the entrance of the house, young and old, were struck blind with a 'blinding light', so that they were helpless to find the door."

It is interesting to note that the angels used some sort of weapon - like a laser beam to blind this perverted mob.

As dawn broke, the angel seized his hand, and the hands of his wife and his two daughters and brought him out and left him "outside the city." Then Lot was told to "flee for your life!"

"As the sun rose upon the earth and Lot entered Zoar, the Lord rained upon Sodom and Gomorrah sulfurous fire from out of heaven." According to the Scriptures.

The most probable location of these cities is beneath the waters of the south end of the Dead Sea, South of the Lisan Peninsula. The waters are very shallow, with an average depth of ten feet. Until recent years, the sea was growing larger because the intake exceeded the rate of evaporation.

In the booklet *Angels,* by Bernard Ward, is an account of how angels sometime appear in order to console or comfort. Country music star Johnny Cash also believes in angels. Three times in his life, beings that Johnny is convinced were angels, came to comfort him, and to let him know that according to God's plan, people close to him were going to die, but that he should not grieve.

When Johnny was a child, an angel appeared to him and told him that his brother Jack would soon die. Two weeks later, his brother was killed in a freak accident.

Another time the angel appeared again to let Johnny know that a close friend would die.

Johnny called his friend's home the next morning, only to find out that he had been killed the night before.

The third encounter came to him the day after his father died. Cash went to bed early that night. He was grieving and depressed. He began to dream:

"I was standing in front of my parents' house," Cash described the dream. "A long, bright car stopped at the curb.
The car had no driver, but the left rear door opened and my father got out and started walking toward me.

"His clear eyes sparkled; they were not covered with the dull film of age I was used to seeing. His teeth were like a young man's, and his hair was full and dark.

"I was waiting for you to come home, I said. I reached out my hand toward him to shake hands. His hand reached out toward mine, and we were only a few paces apart when suddenly a long row of light streamed up from the ground between us."

Johnny wrote that his father smiled and dropped his hand. Then his father spoke:

"Tell your mother that I couldn't come back," the angelic spirit told Johnny. "I'm so happy where I am. I just don't belong here on earth anymore."

Then as Johnny watched, the light grew in intensity and then "all of a sudden he was gone."

Angels protect and assist men who serve God. Sometimes they hinder and oppose men who directly defy God's purposes. Sometimes they destroy men. God has a plan in history and the angels have a role in advancing that plan.

When King Sennacherib of Assyria invaded Judah, his mighty army camped outside the gates of Jerusalem, preparing to invade and destroy the Holy City the next day. But during the night, an angel went through the Assyrian camp and killed 185,000 of the invaders!

The many references to angels in the Old Testament demonstrate that the earliest Hebrews believed that angels were active in their daily lives. An awestruck Gideon proclaimed: "I have seen the angel of the Lord face to face."

Never in Mosaic writings are angels considered mere illusions or figures of speech. The Bible says that angels are an integral part of the story of God's dealing with men. Men of Biblical times recognized the reality of the beings they contacted, and in most cases recognized them as messengers from God.

The prophet Zechariah talked repeatedly with angels who brought him messages from God.

The Bible teaches that God uses angels to work out the destinies of man and nations. He has altered the courses of the busy political and social arenas of our society and directed the destinies of men by angelic visitation many times over. We are told that angels keep in close and vital contact with all that is happening on Earth. Their knowledge of earthly matters exceeds that of men.

In her outstanding book, *Answers From The Angels,* Terry Lynn Taylor recounts a typical child angel encounter as told to her by a mother in North Carolina:

"I have twin daughters named Katie and Amy, who are truly a blessing from God," the mother wrote. "They are 10 years old, and they came to me with a beautiful story of angel help.

Amy said "we were swinging on the swing set out in the backyard, Katie was swinging so high she went way above the edge of the garage roof. Then Katie's foot got caught on the swing bar and pulled her out of the swing way up in the air.

"Katie turned a complete flip in slow motion and landed gently on her bottom in the grass. On each side of her were angels. Where the angels were touching her, on her shoulders
and hips, was a real pretty white light and she landed so softly for being so high in the air."

Joan Webster Anderson, in *Where Angels Walk,* tells another miraculous story of an angel helping a child.

In 1988, Bobby and Debbie Durrance and their sons, Buddy and 12-year-old Mark, moved into a new home in a semi-rural area in Southwest Florida. The subdivision was so new that telephone lines hadn't been installed yet.

One Sunday afternoon, Mark left to play in a recently cleared drainage ditch nearby.

As he leaped across the ditch, Mark felt a vicious pain shoot through his ankle. A rattlesnake had sunk its deadly fangs into his foot. He was more than 150 yards from home, and he could not walk, but he knew in order to get help, he had to get home.

The next thing his mother knew, Mark was calling her from the front porch.

The family was without a phone and 17 miles from the nearest hospital. The parents scooped up Mark who was already going into convulsions, leaped into the family car and raced for help.

All the way to the hospital, Debbie clutched Mark and prayed that his life be spared.

According to Joan Webster Anderson's account: "During the next few days, every part of Mark's body stopped functioning, except his heart. The venom bloated him, swelling his eyes so tightly closed that his lashes were barely visible. His kidneys failed. A respirator moved his lifeless lungs up and down. Internal hemorrhaging caused blood to seep not only from his ears, mouth and eyes, but also from his pores; he required transfusions of 18 pints of blood before the nightmare had ended. There was a 90 percent chance he would lose his leg, and it swelled so large that eventually the doctors slashed it from top to bottom.

Then, miraculously it seemed, Mark awoke from the coma and gradually improved.

Mark's explanation of how he had gotten home, was all he could talk about - he kept saying:

"The man in white helped me. He was just there. When I knew I couldn't make it to the house, he picked me up and carried me. He had on a white robe and his arms were real strong. He reached down and picked me up, and I was hurting so bad that I just sort of leaned my head on him. I felt calm. He talked to me in a deep voice."

"He told me I was going to be real sick, but not to worry. Then he carried me up the stairs to the house and I didn't see him again."

According to Joan Webster Anderson's account, when his mother suggested that the man in white might have been a dream about his grandfather, Mark replied firmly:

"No, mom, that wasn't Grandpa. I didn't dream about the man in white. He was real! He carried me to the house."

Scripture does not tell us what elements make up angels. Nor can modern science, which is only beginning to explore the realm of the unseen, tell us about the constitution of angels.

The Bible seems to indicate that they do not age and never says that one was sick.

Angels are also important figures throughout the New Testament. Gabriel appeared to Zacharias to tell him that he and his elderly wife, Elizabeth, would have a son, John the Baptist, who would be the forerunner to the Messiah.

Gabriel also came to Mary to announce the birth of Jesus. An angel proclaimed the good news of the birth of Jesus to the shepherds of Bethlehem and a host of angels then appeared singing joyful praises.

A guardian angel came to Joseph in the night and warned him to flee into Egypt with Mary and baby Jesus to escape the wrath of King Herod.

An angel of the Lord opened the doors of the prison and let out the apostles after they had been thrown in. Peter was also unchained and let out of prison by an angel.

Though angels are awesome, God forbids us to worship them. And angels consider only the members of the Trinity to be worthy of worship.

How would you live if you knew that you were being watched all the time, not only by your parents, wife or husband or children but also by the heavenly host? The Bible teaches that angels are watching us. Paul says we are a "spectacle" to them.

Angels have watched the drama of this age unfolding. They have seen the church established and expand around the world. They miss nothing as they watch the movement of time.

Angels are not all equal in occupation or power. There are angels who are set as rulers and leaders over other angels, and they occupy a higher place in the work and plan of God.

The Bible teaches that Michael the "archangel" cast Lucifer and his fallen angels out of heaven, and that he enters into conflict with Satan and the evil angels today to destroy their power and to give to God's people the prospect of their ultimate victory. In the mysterious angelic sphere there is obviously even now warfare between God and Satan's angels.

The angel Gabriel appears twice by name in the Old Testament, and twice in the New Testament, each time on errands of the highest consequence in the plan of God.

The prophet Daniel received a vision setting forth future world history in broad outline. When Daniel had seen the vision, and sought to understand it, the angel Gabriel appeared to him, and explained the vision to him. On a different occasion, Daniel said that the angel Gabriel had come to him in "swift flight."

The record of Gideon's commission identifies the one who spoke to him as the angel of Jehovah. We read that the man among the myrtle trees was also the angel of Jehovah, and that Jehovah had sent the horsemen who were to report to this angel.

The ministries of the angel of Jehovah were many and varied. He was obviously God's special representative to his people in the Old Testament, just as Christ was in the New Testament.

This is the angel that disclosed God's name, commissioned Moses to deliver God's people from the Egyptian bondage and to lead them to the Promised Land. He called and commissioned Gideon to go in might against the Midianites. He called and commissioned Sampson through his parents. His protecting power was renown in David's days.

Zachariah pictures the angel of Jehovah as the advocate of God's imperfect believers, defending them against the accusations of Satan. This angel confirmed the covenant with Abraham. The Angel of Jehovah found and comforted the outcast slave woman Hagar.

God promised to send the angel of Jehovah before Moses and Israel to keep them on their journey and to bring them to the Promised Land. They were to obey him and not provoke him.

In the wilderness journey, Moses interceded for Israel after their first breach of the law. God responded by promising: "Behold, my angel shall go before thee." The angel of Jehovah was inside and in control of the "glory cloud" that led Israel in the wilderness. This unique angel guided and protected God's people along their pilgrim pathway.

The second great wave of angels occurred during the middle Ages and was reflected in the art and poetry of the times. The major poets and writers of the time focused on angels.

Later, Emanuel Swedenborg, the renowned Swedish scientist, wrote about his celestial encounters that lasted until his death in 1772. Through these visions, he was able to accurately predict future events.

Swedenborg's writings on angels influenced other great thinkers of his time.

Painting of angels have been done by all the great Renaissance artists. Fra Angelico, the 15th century Florentine artist, saw angels and then painted these personal encounters. Angelico, who considered his work divinely inspired, is considered one of the greatest angel artists of all time.

Joan of Arc, a 16-year-old farm girl, lead an army that defeated the English and saved the French throne. She said she was inspired by the voices of the two archangels, Michael and Gabriel, to take up arms against the invaders.

South pole explorer Sir Ernest Shackleton reported that he and his party were always aware of an angel who traveled with them.

Francis Smythe after he climbed Mt. Everest said; "An angel climbed by my side. I could not feel lonely, neither could I come to any harm. He was always there to sustain me on my solitary climb." Smythe also said that he "divided his meager ration of chocolate to share with the angel."

Charles Lindbergh said an angel "flew with him across the Atlantic and who watched over him when he accidentally dozed off."

In 1975, when Billy Graham published his book, *Angels: God's Secret Agents*, he wrote that he was convinced angels exist to provide unseen aid on our behalf. Billy said "His guardian angel whispered that God himself would welcome Billy when he finally came before Him."

7
Mysteries from the Bible

Noah's Ark: Fact or Fiction?

A Great deluge, which destroyed all life on earth, except for those aboard the Ark, would not quickly be forgotten. This was a catastrophe that affected all mankind. Surely such a destructive universal flood would have been recorded elsewhere besides the Bible. Other historical documents and cultural accounts do support the theory of a great flood.

The Genesis flood was bound to have had a traumatic effect on the survivors - Noah, his three sons and their wives.

Aside from the basic facts and observations of the catastrophe, each survivor no doubt injected his or her own feelings into the story as they told and retold it to each succeeding generations as Noah's sons and wives and descendants repopulated the world.

The story, which has survived about 5,000 years, can be found in some form in more than 200 different cultural accounts of past and present civilizations.

During the reign of Assyrian King Ashurbanipal (627-669) much attention was paid to the records associated with ancient buildings, libraries, and foundation structures. But this recording of historical data was short lived and on the death of King Ashurbanipal, his library in Nineveh containing 100,000 clay tablets was sealed and forgotten until uncovered in the 1850's by a team of British archaeologists.

King Ashurbanipal himself mentions the great flood. He wrote on one clay tablet, "I have read the artistic script of Sumer on the back of Akkadian, which is hard to master. Now I take pleasure in reading of the stones coming from before the great flood."

Versions of the flood tablet have since been found in other places. One early record that combined both the flood and the creation account was the Epic of Atrahasis. This historical tablet was released to the world during the 1960's.

But still the most famous of these accounts is the Epic of Gilgamesh. Three of the Gilgamesh tablets relate the story of Noah's Ark and the Flood using different names and some changes in the details of the event.

The Gilgamesh Epic, written in Akkadian, belonged to the heritage of the great nations of the ancient East. The Epic originated with the Sumerians, the people whose capitol stood on the ancient site of Ur in the present day Iraq. Most Semitic scholars generally agree that the Gilgamesh Epic is older than the Genessis flood account, penned by Moses around 1475 B.C., - about 1400 years after the time of the universal flood.

There are some striking similarities between the Epic and the Biblical account. Merrill F. Unger in his third edition of *Archaeology and the Old Testament, pages 55-56, points them out:*

- Both accounts state that the Flood was divinely planned.
- Both agree that the impending catastrophe was divinely revealed to the hero of the Flood.
- Both connect the Flood with the defection of the human race.
- Both tell of the deliverance of the hero and his family.
- Both assert that the hero was divinely instructed to build a huge boat to preserve life.
- Both indicated the physical causes of the Deluge.
- Both specify the duration of the Deluge.
- Both name the landing place of the boat.
- Each tells of sending birds at certain intervals to determine the decrease of waters.
- Both describe acts of worship by the hero after his deliverance.
- Both allude to special blessings upon the hero after the disaster.

Researcher Alfred M. Rehiwinkel in *The Flood,* on page 162, notes his conclusion: "The two stories are based on the same event, not the same account."

Chinese legends on mainline China say that all Chinese are descendants of "Nu-wah", an ancient ancestor who distinguished himself by overcoming the great Flood.

Dr. E. W. Thwing, a researcher who spent many years in China, has made a successful search into the probable connection between the Chinese story and the Genesis record.

He discovered that ancient Chinese writing has numerous words that can only be traced to "Nu-wah" and the Flood. The word used for "ship" for example, as printed in Chinese books and papers today, is an ancient character made up of the picture of a boat and eight mouths. This shows that the first ship was a boat carrying eight people, Thwing says. The fact that the characters are used for their sound would indicate an ancient man, famous in a great flood, the sound of whose name was kept as "Nu-wah." In the Biblical account that name was Noah. Both are pronounced the same.

While all over the globe a tradition of the great flood may be found, it appears that this is the last great event in which mankind shared. The Bible goes on to mention other amazing events, but none of these subsequent events are commonly found throughout the world as cherished traditions.

There is little doubt that of the more than 200 known cultural flood accounts, the Biblical record is the only one that can really be accepted as an authentic eyewitness account. Sir William Dawson wrote on this subject: "I have long thought that the narrative in Genesis 7 and 8 can be understood only on the supposition that it is a contemporary journal or log of an eye witness incorporated by the author of Genesis in his work. The many details as well as the whole tone of the narrative seem to require this supposition."

Dr. Custance also believes from his research that Genesis is an eyewitness account of the Flood. He writes: "To me it seems almost self-evident that once Noah and his family were inside the Ark and the rains began, from there on God's revelation has not entered into the account. Like any other good captain, Noah kept his daily journal, marking off the events of the days and weeks and months carefully and precisely and accurately, as he and his family experienced them."

Additional evidence of a universal Flood is fossils found atop Mt. Everest, the highest mountain in the world.

The bones of fish and the shells of sea snails and clams have been found atop Everest and other high peaks. Also, geologists who have climbed Mt. Ararat have discovered shells.

Two lakes in the vicinity of Mt. Ararat provide further signs of a deluge.

Mysteries from the Bible

Lake Van, in Eastern Turkey, is 5,640 feet above sea level and yet salty! The Lake Urmia, in Iran is 4900 feet above sea level. It is 90 miles long and 30 miles wide and shallow - only 20 feet deep. Its salt concentration is twenty-three percent.

Many scientists have said the salt lakes remained after the Flood waters had receded. Herring found in Lake Van are a salt-water fish prevalent in the North Atlantic today.

High volcanic mountains surround both Lake Van and Lake Urmia with no outlet to the sea, so they remain salty as they were 5,000 years ago.

Dr. Clifford Burdick, a geologist commissioned by the Turkish government to study Mt. Ararat several years ago, found more impressive evidence of a flood.

He found cube-shaped salt clusters as large as grapefruit near the 7,000-foot elevation of Mt. Ararat, which he says, "indicates that at one time the ocean was thousands of feet higher than it is today.

"As the water dried up, in some places it left little inland seas of water, as that dried up, of course, the salt became more concentrated.

"We don't know of any other time, except during the great flood, that a mountain 17,000 feet high would have been under water. You just have to connect the two. There's no other historical or geological record of the ocean being that high except at that time."

British geologist L. M. Davies, writing about the source of the Flood waters says: "The question as to where the water came from and where it went to will only trouble those who hold extreme views as to the fixity of oceanic and continental levels. If the sea beds can rise and the continents sink, there is no difficulty whatever in finding enough water for a universal flood."

"When we remember that if the whole earth were perfectly flat, the oceans would cover it to a depth of 1.5 miles, this statement is obviously true," says Dr. Frederick A. Filby in his book, *The Flood Reconsidered.* "Either the land sinks or the water level rises - or both."

"Using the metaphors of ancient languages, the Bible records that the windows of heaven were opened, and the fountains of the Great Deep were broken up. In other words, Noah was conscious of torrential rain and the oncoming of a huge tide, not from swollen rivers, but from the Great Deep, that is to say the ocean."

"That the latter (ocean water) was much greater than the former is clearly shown by the fact that the Ark was carried northward towards

Armenia, whereas a river flood resulting from rain would have carried it out to the Persian Gulf," explains Dr. Filby.

Traditionally the search for the Ark's remains is focused on the "mountains of Ararat" since that is where Genesis says it rested 150 days after the Flood began. The "mountains of Ararat" refers to two volcanic summits - Big Ararat and Little Ararat.

Located on the border separating Turkey from the Soviet Union, both history and local legend support the Biblical account that the Ark landed on Mt. Ararat.

The Persians call Mt. Ararat *Kol-I-Nouth*, meaning "mountain of Noah."

The town of Nakhitchevan in 100 A.D. was called Apobaterion, meaning "the landing place."

Nearby Temanin translates to mean "place of the eight," which parallels the number of people on the Ark.

The town of Sharnaskh means "village of Noah." Another village called Tabriz means "the ship."

At the nearby Monastery of Schmiadzin, Armenian priests have a sacred cross, which they say is made from the wood brought from the Ark.

Mt. Ararat's Summit is permanently covered by glacial ice and snow. Seldom is the summit visible because it is obscured by misty, fog-like clouds, which create blizzard conditions most of the year. At lower elevations, thunderstorms are prevalent.

"Because Mt. Ararat is the only glacially covered mountain in that entire region, any evaporation which occurs from the floor or the mountain condenses by mid-afternoon creating the storm conditions," explains Dr. Morris.

The glacial ice cap of Mt. Ararat is more than 200 feet thick, covers about 22 square miles, and begins about the 13,000 to 13,500 foot level. Twelve finger glaciers extend from the main glacier, each with its own name.

Often the sound of thunder can be heard on the mountain. But it is not always thunder. Sometimes it is a chain reaction of echoes created as winds up to 150 miles per hour send gigantic boulders crashing down the Ararat slopes. Recent accounts exist of climbers killed by such boulders, which they could hear but not see as they stood in the mists and fog.

There are absolutely no trees on the mountain for shelter, nor wood to build campfires. Most climbers have suffered from the lack of water because melting snow is quickly absorbed through the porous rocks. At snow-covered elevations, 100-foot deep crevasses often are not noticed until it's too late!

Probably most dangerous are the huge snow and rock avalanches, which climbers set off merely by talking to each other.

History tells us that in ancient times many men who climbed the mountain in search of the Ark returned to report that they had seen it. Accounts from 700 years before the birth of Christ relate the experience of pilgrims climbing Ararat to scrape tar off the sacred vessel from which to fashion good luck omens.

Berossus, the Babylonian high priest from the temple Bel-Merduk, says that in his time, around 300 B.C., remains of the Ark could still be seen and "some get pitch from the ship by scraping it off and use it for amulets."

Hieronymus, an Egyptian historian who authored the ancient history of Phoenicia about 30 B.C. says, "the Ark is there."

Also about 30 B.C., Nicolaus of Damascus, the biographer of Herod the Great, tells of the Ark landing near the summit and states that "relics of the timber were still there."

Nicolaus, wrote a vast universal history from earliest times to the death of Herod. He wrote about the Ark in his 96th book.

Josephus, the famous Jewish historian indicated that in his time, too, there were reliable reports that remains of the Ark could be seen. In about 100 A.D., he wrote in his *Antiquites of the Jews;* "Ark landed on a mountain-top in Armenia. The Armenians call that spot the landing place, for it was there that the Ark came to land, and they show the relics of it even today."

St. Theophitus of Antioch in 180 A.D. wrote of Syrian Antioch, who left his pagan upbringing to become an outstanding apologist for the Christian faith wrote; "The remains of the Ark are to this day to be seen on the Mountains of Ararat."

Encyclopedists of the early middle ages also wrote about the Ark. "Ararat is a mountain in Armenia," writes Isidore of Seille in 600 A.D. He was the first to undertake a compilation of universal knowledge. He goes on to say that historians of his day "testify that the Ark is on Mt. Ararat, and that wood remains of it can still be seen."

Brother Jehan Haithon, a 13th century Armenian prince who became a monk in France, is another who wrote about the Ark. He said that in the snow of Mt. Ararat "one can still see a black spot that is Noah's Ark," which he himself saw in 1254. He writes, "In Armenia there is a very high mountain, and its name is Ararat. On that mountain Noah's Ark landed after the Flood. At the summit a great black object is always visible, which is the Ark of Noah."

Sir James Bryce, a highly regarded British statesman, jurist, and author, made a dramatic and first time solo ascent of the mountain in 1876. He returned with a piece of evidence that should have shocked the scientific world into action.

Braving the hostile environment of Ararat, Bryce uncovered and brought back to London a four-foot long, five-inch thick piece of partially petrified, hand-tooled timber!

He found the wood at 13,000 feet - where no trees had ever grown. He said later in published accounts, "This wood suits all the requirements of the case."

Seven years later in August 1883, the Turkish government announced its discovery of Noah's Ark! It was the first official comment by the Turkish authorities of the Ark's existence.

In 1902, his uncle took George Hagopian, then 10 years old, up the slopes of Mt. Ararat. "It was a year without much snow," he said. "It took us almost eight days to get to the place on the mountain where the holy ship had first come to rest." He climbed on top of the Ark and noticed "the roof was flat with the exception of that narrow, raised section running all the way from the bow to the stern with all those holes in it. I didn't see any nails at all. It seemed that the whole ship was one piece of petrified wood. I could see a green moss growing right on top. No opening of any kind. The other side was inaccessible. I could only see my side and part of the bow."

In 1916, a Russian aviator Vladimir Roskovitsky saw the Ark from the air. "We flew down as far as safety permitted and took several circles around it. When we got close to it, we were surprised to discover the size of it, for it was as long as a city block," said Roskovitsky.

Roskovitsky returned to base and reported his discovery to the captain, who asked to be flown to the site.

"This strange craft," said the captain, "is Noah's Ark!"

When the captain sent the report to the Russian government, it aroused considerable interest, which resulted in an expedition of special soldiers (100 men) going to the site.

Complete measurements of it were taken. The Ark was found to contain hundreds of small rooms; some had fences of great timbers across them. Other rooms were lined with cages.

Only a few days after the report was sent to the Czar, the Bolshevik Revolution overthrew the Russian government.

Eryl Cummings did extensive research of this story to determine its authenticity.

Artists for the last 2,000 years have had many different ideas of what the Ark looked like. The instructions for building the Ark are found in Genesis 6:14-16.

We can say that Noah's Ark was 450 feet long, 75 feet wide, and 45 feet high. It resembled a long rectangular box.

It was the largest wooden vessel ever built. Its roof area was larger than 20 college basketball courts. It would take about 280,000 cubic feet of timber (between 9,000 and 13,000 planks) to construct the boat, which has been estimated at a little more than 4100 tons of dead weight.

The Ark was large enough to have carried more cargo than 569 railroad cars. Or up to 30,000 animals and birds, with lots of room left over for Noah and his family.

Like the giant vessels of today, the Ark's lower center of gravity gave it tremendous stability. The lower it sank in the water due to its heavy cargo, the more stable it became.

George Jefferson Green, an American pipeline and mining engineer working in the Middle East, saw a portion of the Ark in 1952. He was in a helicopter on a reconnaissance mission for his company when he saw the ship protruding from the ice.

Reaching for his camera, Green directed the pilot to maneuver the craft as close as possible to the huge structure below. Green took what turned out to be six extraordinary photographs.

When the photos were enlarged, the joints and parallel horizontal timbers were clearly visible! The photos revealed that the boat was situated on a fault on the mountainside; that a high cliff protected it on one side and a sheer drop-off on the other. About a third of the prow was visible from the air.

Because Ararat is a restricted military zone on the Turkish-Soviet border, the Turks have consistently refused to grant permits allowing the use of aircraft in the fear that planes might accidentally drift across the Soviet border and create an international incident or be shot down by the Russians.

With the birth of the aerospace industry, a new exploration method for Noah's Ark developed - satellites capable of detecting or photographing the Ark on Mt. Ararat. In 1973 Thomas B. Turner, a manager at the McDonnell Douglas Astronautics Company of St. Louis, contacted Dr. John Montgomery to tell him about a satellite photo that possibly showed the Ark.

"Most remarkable was the location, in the very quadrant of the mountain where previous ground sightings of the Ark had been concentrated," says Dr. Montgomery.

Reliable pentagon sources say that these highly sophisticated military spy satellites are capable of photographing the Ark as a well-defined structure.

Our Defense Department source says that military photos of the Ark probably exist in Pentagon files, to which civilians would never have access.

Research indicates that 200 people in 23 separate sightings since 1856 have seen Noah's Ark on Mt. Ararat.

James Irwin, the American Astronaut claims "attempts to remove samples for testing are being blocked by Turkish officials."

Secrets From The Caves

Here is the fabulous story of the Dead Sea Scrolls, perhaps the most momentous news stories of the 20th century. The ancient religious documents whose discovery in a cave in Trans-Jordan, may well be the most important discovery in many centuries.

In 1947, a Bedouin goat man searching for lost animals entered one of the caves high in the cliffs of the Wadi Qumran, about a mile from the Dead Sea. There he stumbled upon several jars somewhat over two feet high, and ten inches wide, containing leather scrolls wrapped in linen clothe.

It was a well-concealed cave in an area of desolation. Its entrance was a hole in a projecting rock, which led into a cave about twenty-five feet in length, and about seven feet wide.

They were removed from the cave and subsequently smuggled to an antique dealer in Bethlehem, and by the end of 1947, the discovery of the manuscripts had become widely known to scholars.

The late E. L. Sukenik, professor of archaeology at the Hebrew University in Jerusalem, recognized the antiquity and the significance of the scrolls, which were at least one thousand years older than any previously known Hebrew manuscript, and dated about 250 years before the time of Christ.

The longest of the manuscripts, is the Isaiah Scroll. It is made of strips of leather, stitched at the edges to form a continuous scroll. It is about a foot wide by twenty-four feet long. Although it shows wear, it has been carefully repaired and its condition may be counted as good. The text is in Hebrew and its fifty-four columns contain the Old Testament book of Isaiah in its entirety! This is the oldest of the Scrolls and the oldest complete manuscript known to be extant of any book of the Bible. The Isaiah scrolls are not further described here since their contents are almost the same as the book of Isaiah in our Bible.

Another manuscript is a *midrash* on the book of Habakkuk. *midrash* is an explanation or commentary applied to a sacred text. The Habakkuk Scroll is only about five feet long and less than six inches wide. Its general condition is good. It could have been written in about the sixth century B.C., and appears to be a warning of the coming invasion by Chaldeans. This manuscript contains the first of the references to a coming Teacher of Righteousness, of which there are six more in this scroll.

It is noteworthy that the Habakkuk Commentary does not include an interpretation of the third chapter of the biblical book of Habakkuk. Even prior to the discovery of the Dead Sea Scrolls many writers had suggested that the third chapter of Habakkuk had been added to the work of the prophet at a latter time. This theory now receives new support.

Another manuscript six feet long, and about ten inches wide, called The Manual Of Discipline, is also in good condition.

The earlier part of the manuscript describes "a Covenant of steadfast love" in which members of a dedicated community are united with God. Then what follows an account of "the two spirits in man," the spirit of light and truth, and its antagonist, the spirit of darkness and error.

Another manuscript called The War Of The Sons Of Light With The Sons Of Darkness, and is very well preserved. This scroll is nine feet in length and six inches wide.

It describes a rather stylized conflict between the righteous and the wicked, and is apocalyptic, like the book of Revelation in the New Testament.

Then come the Thanksgiving Psalms scrolls. There are twenty psalms, very similar to those of the Old Testament.

STRANGER THAN YOU CAN IMAGINE

The last scroll of this collection was called The Lamech scroll before it was opened, because some of the first scholars to see it were able to read the name Lamech on a detached piece of the roll that defied all early attempts to unroll it. When technicians finally succeeded in opening it, they found a scroll about one foot wide and almost ten feet long. It contains chapters from the book of Genesis, expanded and embellished by the introduction of material, which appears to be derived from folklore tradition.

In addition to these major scrolls were fragments of Daniel, which were important because they were written less than 100 years from the original edition of the biblical book of Daniel.

Inscribed fragments of several other biblical books were also found, including Genesis, Leviticus, Deuteronomy, Judges, Samuel, and Ezekiel.

It was late in February 1952, that the archaeologists received word that the Bedouins had discovered another fragment bearing cave at Qumran. Combing an area five miles long on the north side of the Dead Sea, a number of other caves were discovered to contain fragments representing as many as one hundred manuscripts, including Exodus, Jeremiah, and Ruth.

On March 14, a cave with a collapsed roof was excavated with unexpected rewards. This was the cave of the copper scrolls. Lying up against one of the sidewalls were two metal scrolls, one considerably larger than the other. The advanced state of oxidation and the brittle condition of these odd rolls made it impossible for the finders to unroll them immediately. In deed, the problem of opening the copper scroll would perplex the experts for several years.

When unrolled, the copper scrolls listed at least sixty hiding places for gold and silver! These places may be located as far north as Gerizim and as far south as Hebron, a distance of about fifty miles, but the majority of the treasures seem to be buried in the Jerusalem area. Tombs, watchtowers, fortresses, cisterns, and underground passages were the hiding places. Unfortunately these sites are not pinpointed precisely, so it will be exceedingly difficult to use the copper documents as an easy guide to sudden wealth.

According to the copper document, more than 4500 talents of gold and silver were hidden away. One talent is approximately seventy-five pounds, and assumes that our scribe had this weight in mind; the amount of hidden treasure would exceed 170 tons!

The Jewish Zealots who controlled the Jerusalem temple for a short time prior to its destruction by the Romans in A.D. 70 may have prepared the copper document as a reminder of where they had hidden the wealth of the Temple.

In yet another cave, thousands of manuscript fragments were found. In the words of Frank Cross of Harvard, "this collection of fragments is unrivaled in its size and content." More than 380 manuscripts from this cave have been identified. Of this number, about one hundred biblical manuscripts have been identified thus far, including every book in the Hebrew Bible with the one exception of Esther.

The copy of the Genesis Apocryphon discovery at Qumran dates back to the 2nd century B.C., but it was obviously based on much older sources. When discovered in 1947, it had been much mutilated from the ravages of time and humidity. The sheets had become so badly stuck together that years passed before the text was deciphered and made known. When scholars finally made public its content, the document confirmed that celestial beings from the skies had landed on planet Earth. More than that, It told how these beings had mated with Earth woman and had begat giants! (See chapter on giants in this book.)

In the Bible, In Genesis 6; the beauty of the "daughters of man" captivates "The Sons Of God". They subsequently marry them and produce an offspring of giants known as the Nephilim. Genesis goes on to say that these Nephilim were "mighty men and men of renown."

The designation "Sons Of God" is used four times in the Old Testament, each time referring to angels.

In Job 1:6, and 2:1, the "Sons Of God" came to present themselves before the Lord in Heaven. Among the "Sons Of God" is Satan.

Since the designation "Sons Of God" is consistently used in the Old Testament for angels, it is logical to conclude that Genesis 6 is speaking of fallen angels.

Two New Testament passages shed further light on Genesis 6. They are Jude 6-7, and 2 Peter 2:4. These verses indicate that at some point in time a number of angels fell from their pristine state and proceeded to commit a sexual sin.

Jude 6-7 states: "And the angels which kept not their first estate, but left their own habitation, he hath reserved in everlasting chains under darkness unto the judgment of the great day. Even as Sodom and Gomorrah and the cities about them in like manner, giving themselves over to fornication, and going after strange flesh."

These angels transgressed the limits of their own natures to invade a realm of created beings of a different nature. God's law of reproduction is "everything after his own kind."

The book of Isaiah says that the Nephilim and their descendants will not participate in a resurrection, as is the portion of ordinary mortals.

The Nephilim were the superhuman offspring of the bizarre union between the fallen angels and the woman of Earth. They appeared on this planet just before the great Flood. In fact, their existence and vile corruption of the world was the main reason for the catastrophe. Their kind was destroyed along with the rest of mankind in the flood. Only Noah and his family escaped their genetic contamination and hence were saved.

Yet centuries later, the Nephilim emerged again, this time on a more limited scale in the land of Canaan. (See Num. 13:2, 25:33) As before, God ordered their destruction. Why is the presence of the Nephilim so great a threat to man that God would resort to such drastic measures? The reappearance of their progeny would indicate that they returned to earth to again contaminate the human race with their offspring.

Most of the books in what we call the Old Testament were written during a period from about eight to the third centuries B.C., but included fragments that come from documents or inscriptions that are older. A few books are later than the third century, for example, Ecclesiastics.

All of the original manuscripts are lost - although it is now possible that there are fragments of original manuscripts among those recovered from the Dead Sea caves. Even in the first century B.C., only copies were available.

The books of The New Testament were written during a much shorter period: from the last half of the first century to the end of the second century A.D. Here again we have none of the original manuscripts, only copies.

The finding of the Dead Sea Scrolls will serve to reassure us and confirm the general accuracy of the traditional Hebrew text. To be sure, the Isaiah scrolls from the Qumran caves agree with the Masoretic text.

In one of the fragments, a prophecy of the last days included the warning; "in the final battle, God would give victory to the sons of light. The God who led Israel out of Egypt would once more reveal his power, and the enemies would be utterly exterminated. No prisoners of war would be taken. The God who enabled David to smite the mighty Goliath would once more fight for his people and for right."

The Two Men Who Never Died

Enoch is mentioned in the Scriptures, but we are told little about him. Just two verses in the Old Testament, and two verses in the New. These four verses, however, are sufficient to distinguish him as one of the outstanding men of all time. He is listed as the seventh of the ten patriarchs between Adam and Noah. He was the father of Methuselah, the man holding the world's record for longevity. The grandfather of Lamech and great grandfather of Noah, Enoch was known for his exceptional godliness.

The most remarkable fact of all is that he did not die! "Enoch walked with God; and he was not, for God took him." He is one of only two men mentioned in the Bible, and in other material, who were translated to Heaven without tasting death.

In the book of Enoch, other significant facts are given about this patriarch. It claims that Enoch will return to Earth at the end of time, and that he will be one of the two martyrs slain on the streets of Jerusalem.

There can be no question that the clause "and he was not, for God took him" refers to translation; the same expression is used to describe the translation of Elijah, the other man who never died.

After these two men finished their service on Earth, they rose into the sky in a "fiery chariot." Some type of an unidentified flying object, carried both of these men off into space, and they disappeared from the face of the earth!

It seems that Enoch took several flights in what the Bible calls "The Mystery Clouds From Heaven" before his final trip out of this world.

The book of Yasher tells how Enoch would periodically withdraw himself from earthly company, and visit Heaven. After a while he would return to Earth with a divine luster on his face, just as Moses emerged from the presence of God on Mt. Sinai, and his face shone. Enoch obviously had access to knowledge and information completely beyond the reach of mortal man at that stage in man's development. Some of this knowledge Enoch couches in terminology that is allegorical and symbolical. However, there is some amazing data, which we will present, in the next chapter to be called "The Secrets Of Enoch."

It is obvious that Jude had access to the Book of Enoch, and he did not hesitate to quote from it.

Enoch was also given a message to deliver to his young Great Grandson Noah, telling him that the whole of Earth was to be destroyed. He instructed Noah in the way of escape, so that his seed would be preserved for all future generations.

Unquestionably, the Book of Enoch confirms the Biblical record that the earth was defiled and polluted by the incursion of extraterrestrial beings, and particularly by their shameful behavior with the "daughters of man." And as in the Book of Genesis, so in the Book of Enoch, God is incensed with this sexual coupling between celestial (fallen angels) and terrestrial beings, and begins to move in an act of terrible judgment. The Flood, that Enoch had warned Noah of in his youth, some 600 years earlier.

Other books, like The Book Of Jubilee add a few more details to this awesome story. It reveals the date when these heavenly "Watchers" descended to Earth - 461 Annus Mundi, a date, which Bishop Usher would interpret as 3543 B.C. It also notes that these "Watchers" were specifically associated with Jared, the fifth in line from Adam.

Regarding this association with Jared, the Book of Jubilees tells us:

> *"And in the second week of the tenth jubilee of Mahalelel took unto him a wife Dinah, the daughter of Barakel, the daughter of his brother's brother and she bore him a son in the sixth year and he called his name Jared for in his days the angels of the lord descended on the earth, those so named the Watchers."*

Another source of data is the Zadokite Document, which refers to the descent of the Watchers, and to the giant offspring they had with Earth woman.

The Apocalypse of Baruch is another ancient document, which confirms the story of the fallen angels. Written in Syriac, it adds this new concept to the story of the fallen angels: that the source of the corruption was the sinfulness of mankind. It was human sin, and the temptation of "the daughters of man," that caused the angels to fall.

A series of booklets called The Testament of the Twelve Patriarchs, also refer to the fallen angels. They contain nothing like the detail of the Book of Enoch, but they do make an interesting statement: "The woman of Earth were the prime movers in alluring and enticing the angels."

Now these same fallen angels called "the guardians" appear in Enoch's book. God told Enoch to say to the guardians:

> *"Why have ye left the lofty holy place, slept with woman, defiled yourselves with the daughters of men, taken wives unto yourselves and done like the children of Earth and begat sons like giants? Although ye were immortal, ye have defiled yourselves with the blood of woman and produced flesh and blood, as they do who are mortal and perishable."*

Long before this encounter with Enoch, "the most high" had obviously left a crew of his "guardians" on the blue planet Earth and then gone off on other expeditions lasting many years. When he returned, he found to his horror that the "guardians" had had intercourse with the daughters of men.

They were especially trained angels with all the theoretical and practical knowledge for carrying out their mission, but nevertheless, they disobeyed orders by mating!

I suspect that this is the reason that God decided to provide a plan of salvation for mankind. These very angels that corrupted the human race, had been sent to teach and instruct.

Parallels with Enoch from practically every section of the New Testament can be cited. In addition to this apparent literary dependence, however, many of the concepts familiar to us from the New Testament appeared either first or most prominently in Enoch. Thus, for example, the nature of the Messianic reign. Thus also the titles "The Righteous One" the "Elect One," and "The Son of Man." The New Testament concepts of Hell, resurrection and demonology also bear striking similarities to those of Enoch.

Before Enoch disappeared into the cosmos to be with God, he drilled this message into his son:

> *"And now, my son Methuselah, I tell thee all this and write it down for thee; I have revealed everything to thee and handed over to thee the books which concern all these things. My*

son Methuselah, guard these books written by your father and hand them on to the future generations of the world."

Soon after this, God translated him without the experience of death.

Elijah, the other man never to experience death, was a prophet who seemed to be in close contact with some force from another world also.

Once he held a contest on Mount Carmel between Baal and God, and fire fell on Elijah's offering and consumed it.

Elijah told the people "now bring all the people of Israel to Mount Carmel, with all 450 prophets of Baal, and the 400 prophets of Asheran who are supported by Jezebel."

Then Elijah talked to them: "How long are you going to waver between two opinions?" he asked the people. "If the Lord is God, follow him! But if Baal is God, then follow him!"
And I will prepare the other young bull and lay it on the wood on the Lord's altar, with no fire under it. Then pray to your god, and I will pray to the Lord; and the God who answers by sending fire to light the wood is the true God!" And all the people agreed to this test.

So the prophets of Baal prepared one of the young bulls and placed it on the altar; and they called to Baal all morning. But there was no reply. About noontime, Elijah began mocking them.

So they shouted louder and as was their custom, cut themselves with knives and swords until the blood gushed out. They raved all afternoon until the time of the evening sacrifice, but there was no reply, no voice, and no answer.

Then Elijah called to the people. "Come over here." And they all crowded around him as he repaired the altar of the Lord, which had been tore down. Then he dug a trench about three feet wide around the altar.

"Fill four barrels with water," he said, "and pour the water over the carcass and the wood."

After they had done this he said, "Do it again." And they did. "Now do it once more!" And they did; and the water ran off the alter and filled the trench.

Then Elijah prayed aloud, "O Lord God answer me so these people will know that you are God."

Then, suddenly, fire flashed down from Heaven and burned up the young bull, the wood, the stones, the dust, and even evaporated all the water in the ditch! (see Kings 18)

It seems that it was customary for Elijah to disappear in some sort of flying vehicle. For Elisha, a fellow prophet, often ordered a search for Elijah. "It may be that the Spirit of the Lord" has caught him up and cast him upon some mountain or into some valley." He would say.

When Elijah had finished his work on Earth, we read "the Lord was about to take Elijah up to Heaven by a whirlwind". Elisha, who was Elijah's successor, was apparently with Elijah as he was taken away.

"As they were walking along, talking, suddenly a chariot of fire, appeared and drove between them, separating them, and Elijah was carried by a whirlwind into heaven."

Elisha saw it and cried out, "My God! My God! The Chariot of Israel and the Charioteers!"

And as they disappeared from sight he tore his robe!

The "Chariot" is seen as the vehicle by which the angels of God travel, and the "Charioteers" are the angels.

It is important to notice that the "cloud" made an impression on later Hebrew literature. In the Psalms we find that the "Pillar of Cloud" is seen as the vehicle - not God Himself. Here is the link between the "cloud" tradition of Moses and the "Chariot" tradition of Elijah.

The prophet Ezekiel displays an extraordinary gift of observation in describing structure and functions of this type of craft. In his first reported encounter he said:

"I saw a great storm coming toward me from the North, driving before it a huge cloud glowing with fire, with a mass of fire inside that flashed continually; and in the fire there was something that shone like polished brass." (Ez. 1:4 L.B.)

"As I stared at all of this," he said, "I saw four wheels on the ground beneath them." The Chariot had landed!

One of the main themes of the Old Testament religion is "The Exodus," which reported that something resembling a space vehicle and called the "Glory Cloud" by Moses, led the Hebrew people out of Egypt to the Red Sea wilderness, where an "angel" proceeded to give them religious instruction.

"The Lord went before them by day in a pillar of cloud to lead them along the way, and by night in a pillar of fire to give them light, that they might travel by day and by night."

This strange aerial object looked cloud-like during the day and glowed in the dark. Eventually this strange "cloud" seemed to defeat the Egyptians in battle, and it gave guidance and instruction to Moses.

However unlikely as it may seem, the Bible reports that during one of the central events of The Old Testament - The Exodus - some kind of space object was always present, and the Biblical people believed that this object was sent from another world - a world which the Bible calls "The Kingdom of Heaven." The Bible teaches that within this heavenly abode there exists a dazzling, high, and holy city called the New Jerusalem. According to Jesus, this city will be the permanent home for all the redeemed throughout eternity. Both Old and New Testament teaches that this city is suspended in space and will endure forever.

The New Testament focuses on the person of a man named Jesus, who is recorded to have said, *"You are of this world; I am not of this world."* In fact, Jesus often claimed to have come from another world. He is reported to have had contact with beings from another world, such as during His Resurrection from the dead, when "angels" were seen.

Jesus also had close contact on many occasions with what appears to be the same "Glory Cloud" that Moses wrote about: And on at least three occasions the "cloud" proclaimed him to be the "Son Of God" in an audible voice that could be heard by all.

This same "Glory Cloud" appeared to the shepherds at Christ's birth, were seen by several disciples at the "transfiguration" where it cast a shadow, and Moses and Elijah from the Old Testament were "beamed" down to talk with Jesus about His crucifixion, still about two weeks away.

The teachings of Jesus and the events surrounding His life, gains its significance only in the light of the climax. If Jesus was an enigma to the religious leaders and to the common people of His day, it was because He had come to fulfill a unique mission, which found its fulfillment in an act of crucifixion.

After the reported resurrection of Jesus from the dead, Luke records the "ascension" of Jesus into the "talking" - "bright" - "Glory Cloud."

It took place at Bethany when after lifting up His hands to bless the disciples, "Jesus vanished from sight" (Luke 24:50.51)

However, the first chapter of Acts fills in the picture with more detail:

"And when he had spoken these things, while they beheld, He was taken up; and a cloud received Him out of their sight. And while they looked steadfastly toward heaven as He went up, behold, two men (angels) stood by them in white apparel; which also said, 'Ye men of Galilee, why stand ye gazing up into heaven? This same Jesus, which is taken up from you into heaven, shall so come in like manner as ye have seen Him go."

All four gospels suggest that there were unusual persons - beings from another world - present at the empty tomb. These same beings were now present as Jesus left the Earth in the "Glory Cloud."

One of the key turning points in the history of the church was the conversion of Paul the Apostle. It was while the risen Christ confronted Paul traveling to Damascus to extradite Christians.

At mid-day a light from the sky flashed about him and his band of men, throwing them all to the ground and blinding Paul. Then a voice from the "Mystery Cloud" was heard to say, "Why do you persecute me?" Paul asked the identity of the speaker and was told, "I am Jesus whom you are persecuting." He was then instructed to rise and enter the city, and he would be told what to do.

Paul was later instructed that he had been appointed by Jesus to be an Apostle to the Gentiles, delivering to them the message of a crucified and risen Lord and bringing them into the unity of one body in Christ.

The Bible also predicts that the "Clouds of Heaven" will appear at the rapture of the church, and indeed, will be the vehicles that will take the believers from the Earth "beaming" them up into the "clouds." The Bible also says that these same "Mystery Clouds" will be present during the great tribulation period, and at the second coming of Christ to the earth.

Are angels simply part of our inherited religious mythology, or were superior beings from another world really an important force behind the Biblical religion?

Angels are mentioned in thirty-four books of the Bible a total of 273 times.

The Bible clearly claims in many key instances that "Unidentified Flying Objects" played a significant role in the development of the Hebrew-Christian faith, and it also claims that superior beings from another world made significant contributions at various times.

The Secrets Of Enoch

Capernicus wrote his main work *Six Books On The Revolutions Of the Celestial Bodies* in 1534, Galileo Galilei discovered the phases of Venus and Jupiter's moon in 1610 with a homemade telescope. In 1609, Johannes Kepler discovered the two laws of planetary motion, for which he was the first person to give a dynamic explanation.

But Enoch had this information in 3543 B.C., and wrote it down in a book called the *Secrets Of Enoch.*

According to Moses, Enoch was a pre-flood prophet, the son of Jared. His son was Methuselah, and his great grand son was Noah.

After his service on earth, the prophet Enoch rose into the heavens in a "fiery chariot." But he left this book behind, entrusted to his son Methuselah, who was to give the information to future generations. (For the men of his time were not able to understand the technical connotations, which were aimed at a later age.)

"On orders from the most high," Enoch took down the data literally so that it would be understood in ages to come. His compendium on astronomy, complicated fractions and exponential series (which are incredibly close to our own mathematical knowledge) cover many pages, and give us deep insights into the earliest mysteries of astronomy. Where is this prophet supposed to have gotten these data if not from "angels?"

The early Abyssinian church had accepted Enoch's scriptures in their canon. But it was not until 1773 that the African explorer J. Bruce brought a copy back to England. The book was translated into German for the first time at Frankfurt in 1885. In the meantime, fragments of a very early Greek transcript were discovered. Comparison of the Ethiopian and Greek texts showed that they tallied, so we can assume that we now possess an authentic copy of this book.

Moreover, fragments of manuscripts found at Qumran, called *The Dead Sea Scrolls,* also authenticate these writings.

Although not known in Europe until the 18th century, the book was a veritable best seller in the days of Jesus. In the centuries immediately preceding and following his birth, this book was widely read and discussed, and its impact was tremendous. Without question it is the most notable apocalyptic work outside the canonical Scriptures. R. H. C. Charles, an outstanding authority in his field, tells us "the influence of Enoch on the New Testament has been greater than that of all the other apocryphal books put together.

For one thing, the book of Enoch gave the world the concept of a pre-existent Messiah, and in doing so prepared the way for Christian doctrine. It was from this same book that the *Manual of Discipline* (also found at Qumran) received its solar calendar. What is more, this book became so influential that it became an exemplar and a catalyst for the burgeoning apocalyptic literature of the time. Indeed, it can be claimed, the Book Of Enoch was one of the most important books ever written.

Tertullian and some of the other Church Fathers considered it of such import that they included it as part of the sacred canon of Scripture.

There is also in the book of Enoch a doctrine that one finds nowhere else. We are told that each nation has its own "angel shepherd," and that Israel's shepherd till the last years of the kingdom of Judah; was God Himself. Then in disgust He turned them over, not to their own guardian, but to the angel, who was over-seeing, the angels of the Gentiles.

Such a doctrine prefigures and predicts the "times of the Gentiles" mentioned in the Bible. Yet, of all mysteries, this book remained neglected for thousands of years, and continues to remain neglected in the 20th century.

In this book, Enoch announces a last judgment. He claims "the divine God will leave his heavenly abode to appear on earth with his angelic host." He describes the fall of the rebel angels; He gives the names of the angels who couple with the daughters of man against the orders of God; Enoch describes his travels to different worlds and distant "firmaments." The book contains all kinds of parables, which he says God told him. Enoch gives an astonishing accurate detail of the orbits of sun and moon, and the stars and functioning of the heavens. For example he writes: "I saw lightning and the stars of heaven, and how they were all named by their names and weighed with a genuine measure, according to the intensity of their light, the breadth of their places and the day of their appearance."

It is a fact that astronomers classify stars by their names and their order of magnitude, (weighed with a genuine measure) their brightness, (intensity of their light) their position, (breadth of their places) and the day when they were first observed (day of their appearance).

He goes on to say, "the thunder has fixed laws for the duration of the clap which is allotted to it. Thunder and lightning are never separated."

As we know, thunder is caused by the sudden expansion of air heated by lightning and moves at the speed of sound (333 meters a second.) Thunder does have fixed laws for the duration of the clap. How much earlier would natural laws have been discovered if texts like this had been scrutinized?

In one of Enoch's trips from Earth to what must have been some type of command module, which was in orbit around the Earth, he describes the gleaming metal hull of the "mystery cloud from Heaven," as "built of crystal stones." (The idea of angels traveling in some type of space vehicle is also mentioned many times in both the Old and New Testament. The prophet Daniel in (7:13) also uses the phrase "clouds of heaven." Other descriptions found in the Bible are: "The pillar of cloud," "the bright cloud," "the glory cloud," "heavenly clouds," "the glory of the Lord," and several others.)

Enoch also saw the dazzlingly bright spaceship wall on the side facing the sun. (Or did the blinding jet exhausts of breaking rockets astound him?) Undoubtedly he was afraid of having to step into the fire. He is all the more surprised a moment later to find that the interior of the "house" is cool.

The "guardians" have brought Enoch to this place, and requested that he intercede for them with "the most high."

And Enoch writes:

"I heard the voice of the most high; 'Fear thou not Enoch, thou righteous man and scribe of righteousness. Go thou and speak to the guardians of Heaven who have sent thee in order to intercede for them. For they should really intercede for man, and not men for them.'"

Who are these "guardians of heaven," Ezekiel mentions these strange figures; they appear in the Epic of Gilgamesh; and they haunt the fragmentary texts of the mysterious Lamech scrolls, which were found in caves high above the Dead Sea.

There is no doubt that they are fallen angels!

The book of Enoch suggests that the fall of angels occurred because of the marriage of "the sons of God" with the "daughters of man." The angels in turn taught mankind the various arts and skills of civilization, and mankind became corrupted and godless under the direction of the "watchers." R. H. C. Charles translates the word "watchers" as angels, or children of heaven.

The Book of Enoch alleges that two hundred of these "watchers" descended to Earth in the days of Jared. Some are named. The worst one of all is called Azazel. The name occurs in other Jewish documents, like the Apocalypse of Abraham. Azazel is accused of having "scattered over the earth the secrets of heaven and hath rebelled against the Mighty One."

The book of Jubilees also says that the Watchers came to Earth in order to "instruct the children of men and bring about justice and equity on the Earth." However, the story ends the same way: instead of instigating justice and equity, they lusted after the woman of Earth, and merited the full judgment of God. Which would come some 600 years later.

According to the Book Of Enoch, these Watchers instructed the people of Earth in many studies, including the arts of magic, astrology, and how to fashion weapons of war and destruction, particularly swords, knives, and shields.

The book of Enoch confirms the book of Genesis to the letter when it states: "There arose much godlessness, and they committed fornication, and they were led astray, and became corrupt in all their ways."

Things got so bad on planet Earth, according to the Book of Enoch, that the archangels of heaven - Michael, Uriel, and Gabriel - reported the matter to the most high.

God was moved to anger at these watchers, because of the horrible practices that they had introduced upon the Earth. And there was something else: God was angry at the fact that they had disclosed certain secrets, and were teaching them to their sons, the giants. "We are not told exactly what these secrets were," writes I. D. E. Thomas, "except that they were eternal secrets which men of Earth were striving to learn, and Which God did not intend for fallen man to discover."

Enoch is instructed by "the most high" to deliver this warning to the watchers:

> *"Enoch, thou scribe of righteousness, go, declare to the Watchers of heaven, who have left the high heavens, the holy eternal place, and have defiled themselves with woman, and have done as the children of earth do, and have taken unto themselves*

> *wives: Ye have wrought great destruction on the earth: And ye shall have no peace nor forgiveness of sin."*

These same watchers in turn approached Enoch to mediate on their behalf, and to write out a petition in their favor. God, however, rejects the petition and Enoch is summoned to speak to the watchers again:

> *"Go, say to the watchers of heaven, who have sent thee to intercede for them: "Ye should intercede for man, and not men for you. Wherefore have ye left high and holy and eternal heaven, and lain with woman and defiled yourselves with the daughters of men, and taken wives unto yourselves and done like the children of earth and begotten giants as sons. And although ye were holy, spiritual living and eternal life, you have defiled yourselves with the blood of women, and have begotten children with the blood of flesh, have lusted after flesh and blood as those who do die and perish."*

And as for the giants or Nephilim produced by them:

> *"The giants, who are produced from the spirits and flesh, shall be called evil spirits upon the Earth. Evil Spirits have proceeded from their bodies; because they are born from men and from the holy watchers is their beginning and primal origin. And the spirits of the giants afflict, oppress, attack, do battle, and work destruction on the earth. And these spirits shall rise up against the children of men and against the woman, because they have proceeded from them. Thus shall they destroy until the day of the consummation, the great judgment in which the age shall be consummated over the Watchers and the godless, ye, shall be wholly consummated. And now as to the watchers, say to them therefore: Ye shall have no peace."*

Such was the message given to Enoch, who transmitted it to the Watchers and their progeny on Earth, the giants.

Enoch was also given a message for his great grand son Noah, telling him that the whole Earth was to be destroyed. He was also to instruct Noah in the way to build the ark in which he could escape, so that his seed is preserved for all future generations. Many years later he did just that.

I.D.E. Thomas writes: "Unquestionable, the book of Enoch confirms the Biblical record that the Earth was defiled and polluted by the incursion of extraterrestrial beings, and particularly by their shameful behavior with the "daughters of man." And as in the Book Of Genesis, so in the Book of Enoch, God is incensed with this sexual coupling between celestial and terrestrial beings, and begins to move in an act of judgment." And the Lord said. I will destroy man whom I have created from the face of the Earth.'"

We also find reports of fallen angels in many other cultures and among many other nations.

Most people are acquainted with the mythologies of ancient Greece and Rome. The gods or semi-gods in these traditions go under many different names, but the behavior has a common denominator. Whether these gods are called Zeus or Jupiter, Poseidon, or Neptune, Aphrodite or Venus, Eros or Cupid, their sex orgies, promiscuity's, cruelties and violence are all of the same cloth. And so are their offspring.

The Genesis story, according to Tom Horner, corresponds precisely to the age of the Heroes in ancient Greece. These heroes were also "spawned by divine fathers" and human mothers! One of them was Hercules.

The story of Zeus also is well known. Promotheus was aware of the secret that Zeus had no control over his lusts, and aware also of the names of the woman whom he would seduce.   Eventually however, he and Zeus were reconciled. But cruelty was not the only distinctive of Zeus. There seemed to be no boundaries or limits to his lust. Emile Gaverluk writes:

> *"Zeus' amorous victories illustrate the actions of uncontrolled spirit-beings lusting after human flesh. The Whole story of Greek mythology is an expanded version of that astonishing verse in the Bible; "The sons of God saw the daughters of men that they were fair; and took them wives of all they chose." The mythology of the past is a startling revelation of the uncontrolled behavior of both spirit-beings and rebellious man."*

From India come the Mahabharata and other ancient Sanskrit texts, which tell of "gods" begetting children with women of earth, and how these children inherited the "supernatural" skills and learning from their fathers.

Similar stories are found in the Epic of Gilgamesh, where we read of "Watchers" from outer space coming to planet Earth, and producing giants.

I. D. E. Thomas writes: "that when these and other accounts, are all tied together, they amaze us by their common core. Each one refers, with slight variations, to the traffic between "the sons of god" and "the daughters of man," and the sexual activities in which they engaged, and to the unusual and abnormal offspring they produced."

Although Enoch had much to say about the fallen angels, who couple with the daughters of man, and produced giant offspring, he also had many revelations concerning the creation, history of mankind, the future of nations, and the world, and the time at the end of history.

Long before man had the telescope - or any knowledge about sub-atomic particles, Enoch wrote: "All things that are seen, are created out of things that can not be seen!" And Enoch refers to "creatures both visible and invisible!

He predicted the future history of Israel; He predicted the flood as a judgment upon the world; He recounts the history from the beginning, to the time of Enoch. He goes on to predict the history of the world to the founding of the Messianic Kingdom; He predicts the increase of sin after the flood until the time of the 1,000-year Messianic reign; He writes:

"Just as God performed His creative work in six days and rested the seventh, even so the history of the world would span 6,000 years from Adam, and it would then rest for 1,000 years. After this, an eternal day of blessing would begin."

He tells us "that the souls of men were created before the world began and also a place either in Heaven or Hell for the future habitation of each soul. The soul was created good, but because of the body, sin appeared in spite of the instruction man had received regarding the two ways."

Then the prophet Enoch, his work on Earth completed, rose into the heavens in a fiery chariot - one of the mystery clouds from heaven - and disappeared into the sky.

Noah carried a copy of these records, on board the great ship that he built, in order that they would survive the flood, and be available to all future generations.

Mathematical Proof of the Bible's Authenticity

If you write a document and send it via e-mail, how do you know the message arrived at its destination uncorrupted by error? In the computer age, it is simple. You just need to create a "checksum" for the file you are sending. If the checksum of the received file does not match that of the sent file, then you know something has gone wrong.

What most people do not realize is that every packet of information sent via the Internet has a checksum, the same as each and every sector on the hard drive of your PC. Checksums are those invisible attributes, which monitor the integrity of nearly every operation you perform on your computer.

Checksums count the values of each character in a computer file so the file can later be checked to ensure that it conforms to its original checksum. Variances do exist in implementations but the purpose is the same. In the computer age, checksums are the unseen guardians of data integrity.

More than 3,500 years ago, the luxury of computers did not exist so what could you have done if you planned to write a document, which had to be completely error free. Your only option was to use known mathematical markers in the hope that someone down the line would pick up on them and recognize them. You could perhaps use several different schemes concurrently just to prove without a doubt that the content was correct at a later date.

Hebrew and Greek are both alphanumeric languages. Therefore, these languages used various letters of their alphabet to also express numbers. Each letter in these languages stood for a number. In other words, each letter expressed both a letter and a number. Every single word in Hebrew contains a series of letters with individual numeric values.

When we examine the Hebrew text of Genesis 1:1 we discover the built in "checksums;" an incredible phenomenon of **exact multiples of seven** that could not be there by chance. There are 30 separate codes involving the number 7 in this first verse of the Bible alone. For example:

The number of Hebrew words = 7.

The number of letters equals 28, or 4 sevens.

The first three words contain 14 letters, or 2 sevens.

The last four words have 14 letters, or 2 sevens.

The fourth and fifth words have 7 letters. The 3 key words *God, Heaven, Earth,* have 14 letters, or 2 sevens.
    The number of letters in the four remaining words is also 14, or 2 sevens.
    The middle word is the shortest with 2 letters. However, in combination with the word to the right or left also total 7 letters.
    The numeric value of the first, middle, and last letters is 133, or 19 sevens.
    The numeric value of the first and last letters of all seven words is 1,393 or 199 sevens.
    Professors on the mathematics faculty at Harvard University were unable to duplicate this incredible mathematical phenomenon. The Harvard scientist used the English language and artificially assigned numeric values to the English alphabet. They had a potential vocabulary of over 400,000 available words to choose from to construct a sentence about any topic they chose. (Compare this to the 4500 available word choices that the writers of the Old Testament could use.) Despite their advanced mathematical abilities and access to computers, they were unable to come close to incorporating 30 mathematical multiples of 7 as found in the Hebrew words of Genesis 1:1.
    (Throughout the Bible the number seven appears repeatedly as a symbol of divine perfection – the seven days of creation, God rested on the 7th day, the 7 churches, the 7 seals, the 7 trumpets, etc. The number 7 appears 287 times, (or 41 sevens) in the Old Testament. The word seventh occurs 98 times, (or 14 sevens.) The word seven-fold appears 7 times. The word seventy is used 56 times, (or 8 sevens.) These mathematical patterns underlie all of the books of the Old Testament.)
    It is interesting to note that all 5 books of the Torah also incorporate the mathematical properties of "PI" in the word counts of each book.

What about the New Testament? Is there "checksums" used in the Greek also to validate the integrity of those scriptures?

Yes! We have precisely the same kind of evidence that proved the Hebrew Old Testament to be inspired.

The first 17 verses of the New Testament contain the genealogy of Christ. It consists of two main parts: Verses 1-11 cover the period from Abraham, the father of the chosen people, to the Captivity, when they ceased as an independent people. Verses 12-17 cover the period from the Captivity to the promised Deliverer, the Christ.

This 17-verse text contains 72 Greek Vocabulary words. (A Vocabulary word is a particular word different from any other word that was used in a passage.) In other words, a Greek word that appears 4 times in the passage would only be counted as 1 vocabulary word. The same system of numeric and place values for letters exists in the Greek language used in the New Testament as we found in Hebrew of the Old Testament.

A listing of the phenomenal features of sevens found in Matthew 1:1-17 are as follows: The total numeric value of the 72 vocabulary words is 42,364, (or 6,052 sevens.) The number of Greek nouns in this passage is 56, (or 8 sevens.)

The Greek article for "the" occurs 56 times, (or 8 sevens.) In the first eleven verses (Matthew 1:1 – 11) we find these additional features:

The number of Greek vocabulary words is 49, (or 7 sevens.) Of these 49 words, 28 begin with a vowel, (or 4 sevens.) Of these 49 words, 21 begin with a consonant, (or 3 sevens.) The numeric value of these 49 words equals 266, (or 38 sevens.) Of these 49 words, 35 words occur more than once, (or 5 sevens.) Of these 49 words, 14 words occur only once, (or 2 sevens.) The number of proper nouns is 35, (or 5 sevens.)

The number of times these proper nouns occur is 63, (or 9 sevens.) Of the 35 proper nouns, the number of male names is 28, (or 4 sevens.) Three women, (Tamar, Rahab, and Ruth,) are named in this section. The number of Greek letters in these three names is 14, (or 2 sevens.)

These were only a few of the numerical features that are found in this passage. To illustrate that this is not an unusual phenomenon occurring only in this one isolated passage, I will list some additional mathematical features that occur in the balance of Matthew

1:18-25. In this seven-verse section God has placed an astonishing pattern of SEVENS that verify the signature of the original author.

The number of vocabulary words is 77, (or 11 sevens.)

Six Greek words occur only in this passage and never again in Matthew. These six Greek words contain precisely 56 letters, (or 8 sevens.)

The number of distinct proper nouns in this passage = 7

The number of Greek letters in these 7 proper nouns is 42, (or 6 sevens.)

The number of words spoken by the angel to Joseph is 28, (or 4 sevens.)

The number of Greek forms of word used in this passage 161, (or 23 sevens.)

The number of Greek forms of words in the angels' speech is 35, (or 5 sevens.) The number of letters in the angel's 35 forms of words is 168, (or 24 sevens.)

Now let us examine the first part of this genealogy alone: Its vocabulary has 49 words, (or 7 sevens.) This number is itself seven sevens, and the sum of its factors is 2 sevens. Of these 49 words, 28 (or 4 sevens) begin with a vowel; and 21, (or 3 sevens,) begin with a consonant. These 49 words of the vocabulary have 266 letters, (or 38 sevens.) The sum of its factors is 28 or 4 sevens, while the sum of its figures is 14 or 2 sevens. Of these 266 letters, 140 or 20 sevens are vowels, and 126 or 18 sevens, are consonants. That is to say; Just as the number of words in the vocabulary is a multiple of seven, so is the number of its letters a multiple of seven; just as the sum of its factors of the number of the words is a multiple of seven, so is the sum of the factors of the number of their letters a multiple of seven. And just as seven divides the number of words between vowel words and consonant words, so is their number of letters divided between vowels and consonants by sevens. Again: Of these 49 words, 35 or 5 sevens occur more than once in the passage, and 14 or 2 sevens, occur but once. Seven occurs in more than one form, and 42 or 6 sevens, occur in only one form. And among the parts of speech the 49 words are thus divided:

42 or 6 sevens are nouns, 7 are not nouns. Of the nouns, 35 or 5 sevens are Proper names; seven are common nouns. Of the Proper names 28 or 4 sevens are male ancestors of the Christ, and seven are not. Moreover, these 49 words are distributed alphabetically thus words under 'alpha - epsilon' are 21 in number, or 3 sevens; 'sigma - iota' 14, or 2 sevens; 'lambda - upsilon' also 14 or 2 sevens. No other groups of sevens stopping at the end of a letter are made by these 49 words, the groups of sevens stop with these letters and no others. But the letters, alpha, epsilon, stigma, iota, lambda, upsilon, are letters 1, 5, 6, 10, 12, 22 of the Greek alphabet, and the sum of these number (called their Place Values) is 56, or 8 sevens. This enumeration of the numeric phenomena of these 11 verses does not begin to be exhaustive, but enough has been shown to make it clear that this part of the genealogy is constructed on an elaborate design of sevens.

Let us now turn to the genealogy as a whole once again. I will not weary you with recounting all the numeric phenomena thereof. I will point out only one feature. The New Testament is written in Greek. The Greeks had no separate symbols for expressing numbers, corresponding to our Arabic figures, but used instead the letters of their alphabet just as the Hebrews, in whose tongue the Old Testament is written, made use for the same purpose of theirs. Accordingly, the Greek letters stand for the following numbers: 1, 2, 3, 4, 5, 6, 7, 8, 9, 10, 20, 30, 40, 50, 60, 70, 80, 90, 100, 200, 300, 400, 500, 600, 700, 800, 900.

Every Greek word is thus a sum in arithmetic obtained by adding the numbers for which its letters stand, or their numeric values. Now the vocabulary to the entire genealogy has 72 words. If we write its numeric value over each of these 72 words and add them, we get for their sum 42,364, or 6,052 sevens, distributed into the following alphabetical groups only: alpha - beta have 9,821 or 1,403 sevens; gamma - delta, 1,904 or 272 sevens; epsilon - stigma, 3,703 or 529 sevens; theta - rho, 19,264 or 2,752 sevens; sigma - chi, 7,672 or 1,096 sevens. But the numeric value of the 10 letters used for making these groups is 931 or 7 x 7 x 19, a multiple not only of seven but also of seven sevens.

The second part of this chapter, verses 18-25, relates the birth of the Christ. It consists of 161 words, or 23 sevens; occurring in 105 forms, or 15 sevens, with a vocabulary of 77 words or 11 sevens. Joseph is spoken to here by an angel. Accordingly, of the 77 words the angel uses 28 or 4 sevens; of the 105 forms he uses 35 or 5 sevens; the numeric value of the vocabulary is 52,605 or 7,515 sevens; of the forms, 65,429 or 9,347 sevens. This enumeration only begins, as it was barely to scratch the surface of the numerics of this passage. But what is especially noteworthy here is the fact that the angel's speech has also a scheme of sevens making it a kind of ring within a ring, a wheel within a wheel.

The second chapter of Matthew tells of the childhood of the Christ. Its vocabulary has 161 words, or 23 sevens, with 896 letters, or 128 sevens, and 238 forms, or 34 sevens; the numeric value of the vocabulary is 123,529 or 17, 647 sevens; of the forms, 166,985 or 23, 885 sevens; and so on through pages of enumeration. This chapter has at least four logical divisions, and each division shows alone the same phenomena found in the chapter as a whole. Thus the first six verses have a vocabulary of 56 words, or 8 sevens, etc. There are some speeches here: Herod speaks, the Magi speak, and the angel speaks. But so pronounced are numeric phenomena here, that though there are as it were numerous rings within rings, and wheels within wheels, each is perfect in itself through forming all the while only part of the rest.

There is not, however, a single paragraph of the scores in Matthew that is not constructed in exactly the same manner. Only with each additional paragraph the difficulty of constructing it increases not only in arithmetical, but in geometrical progression. For he contrives to write his paragraphs so as to develop constantly fixed numeric relations to what goes before and after. **Thus in his last chapter he contrives to use just 7 words not used by him before.**

A second fact is yet more important: In his very first section, the genealogy discussed above, the words found nowhere else in the New Testament occur 42 times, or 6 sevens, and have 126 letters, or 18 sevens. Each number a multiple not only of seven, but of 6 sevens, to name only two of the many numeric features of these words. **But how did Matthew know, when designing this scheme for these words (whose sole characteristic is that they are found nowhere else in the New Testament) that they would not be found in the other 26 books? That the other 7 New Testament writers would not use them? Unless we assume the impossible hypothesis that he had an agreement with them to that effect, he must have had the rest of the New Testament before him when he wrote his book. The Gospel of Matthew, then, must have been written last (but read on.)**

It so happens, however, that the Gospel of Mark shows the very same phenomena. The Last Twelve Verses of Mark, presents among some sixty features of sevens the following phenomena: It has 175 words, or 25 sevens, a vocabulary of 98 words, or 2 sevens of sevens, with 553 letters, or 79 sevens, 133 forms, or 19 sevens, and so on to the minutest detail. Mark then, is another miracle, another unparallel literary genius. And in the same way in which it was shown that Matthew wrote last it is also shown that Mark, too, must have wrote last. Thus to take an example from this very passage: It has just one word found nowhere else in the New Testament, θανάσιυος, meaning "deadly." This fact is signaled by no less than seven features of sevens, thus: its numeric value is 581 or 83 sevens, with the sum of its figures 14, or 2 sevens, of which the letters 3, 5, 7, 9 from the beginning of the word have 490, or 7 x 7 x 5 x 2: a multiple of seven sevens, with the sum of its factors 21, or 3 sevens. In the vocabulary it is preceded by 42 words, or 6 sevens. In the passage itself by 126 words, or 7 x 6 x 3, both numbers multiples not only of seven, but of 6 sevens. We have thus established before us this third fact to contemplate: **Matthew surely wrote after Mark, and Mark just as surely must have written after Matthew. (But read on.)**

It happens, however, to be a fourth fact that Luke presents the same phenomena as Matthew and Mark, and so does John, and James, and Peter, and Jude, and Paul. And we have thus no longer two great unheard-of mathematical literati, but eight of them and each must have written after the other.

And not only this: as Luke and Peter each wrote 2 books, John 5, and Paul 14, it can in the same way be shown that each of the 27 New Testament books must have been written last.

The phenomena are there and there is no human way of explaining them. Eight men cannot each write last, 27 books, some 500 pages, cannot each be written last. But lets assume that one Mind directed the whole, and the problem is solved simply enough: by the words of this Verbal Inspiration – all Scripture is the word of God.

Just as the number seven is found repeatedly hidden beneath the text of Scripture we find the same number used repeatedly by the Creator in His physical creation of the universe and it's inhabitants. There are precisely seven colors in the light spectrum that merge together to form light. The study of music again reveals that there are exactly seven musically whole tones in the scale. The whole human body is renewed at the cellular level every seven years. The gestation cycle to produce a baby takes 280 days, or 40 sevens. We are commanded to rest every seventh day. The Bible says our life expectancy is seventy years or 10 sevens. Almost all animals have gestation periods that are multiples of seven: Lion 98 days or 14 sevens; Sheep 147 days or 21 sevens: Hens 21 days or 3 sevens; Ducks 28 days or 4 sevens; Cats 56 days 8 sevens; Dogs 63 days or 9 sevens, Etc.

So we know that whatever codes are in the Bible, God Himself put them there (but only in the original language, as they are lost with any and all translations).

8

The Inexplicably Bizarre

Lives Saved by Animals

In 1976 a group of scientists, who were gathered to study earthquakes at the Center for Earthquake Research in Menlo Park, California, concluded that all manner of animals exhibit unusual behavior just prior to impending disaster.

Animals seem to know long before humans were alerted to a danger. In fact the Chinese have been able to forecast earthquakes using animal behavior as a forewarning system.

Shoo-Shoo is a 10-year-old Siamese cat who lives at the ABLE Inc. Group Home for the Developmentally Disabled in Dickinson, North Dakota. The cat knows in advance when someone is about to have an epileptic seizure. The cat begins to meow and cry, circling continuously around the patient's wheelchair or bed. The cat refuses to leave the patient's side until a member of the staff arrives.

Mardi Hadfield of Tucson, Arizona, has a cat that also warns her of an epileptic attack. The cat paws repeatedly at her leg five minutes or so before the seizure.

Andrew Rowan, director for Animals and Public Policy at Tufts School of Veterinary Medicine in Frafton, Mass., says many dogs have the ability to warn owners of approaching seizures also.

Time and again, animals have demonstrated uncanny senses far beyond our ability to understand, and they have used these strange powers to save human lives.

In 1980, a seagull saved a woman's life. The circumstances, attested to by a number of responsible people, are as follows:

An 82-year-old Cape Cod resident, Rachel Flynn, slipped and fell more than thirty-five feet, and landed on a deserted section of beach, trapped between large boulders, and badly hurt. She thought she was going to die.

But then, a seagull, that Mrs. Flynn had been feeding regularly, landed nearby, and seemed to act erratically.

The gull then flew for more than a mile to Mrs. Flynn's home, and began pecking continually on a window with it's beak, and flapping its wings frantically.

Rachel's sister June was at first irritated by all the noise. But after about fifteen minutes she decided that the bird was trying to tell her something.

June began to follow the bird, which stopped every few feet, to make sure that the elderly woman could keep up. The bird stopped at the cliff's edge, and when June looked over, she could see her sister lying below.

There was no doubt in the sister's mind that the seagull had intentionally gone for help, and in doing so, Mrs. Flynn's life had been saved.

In 1977, a 15 year old boy, Kirsten Hicks left his Persian cat Howie with his grandparents, who lived a thousand miles away from the boy. Soon the Grandparents reported to Kirsten that the cat had vanished, and all efforts to find it had failed.

A year later the cat, with bloody paws, arrived at Kirsten Hick's home! The cat had made a twelve-month trek across vast tracks of wilderness, across rivers, and deserts, and over many miles of rugged terrain, to somehow find its home. How the cat could accomplish this feat, when all of the area covered was totally unknown to it, remains an unsolved mystery.

In a similar case, Jean-Marie Valembois left his sheep dog Black with a cousin some 500 miles away, because of his new job at a construction site. He was staying at a hotel, which would not admit pets. But then one day, several weeks later, the dog showed up at the construction site with bloody paws and whining. How could this dog have made his way to a place he had never been before? Or for that matter, how did Black know his master was working at this place?

In 1983, Oscar Simonet, a three-year-old boy at the time, had disappeared on the rugged cliffs near his home in Villacarlos, Menorca. Search parties could not find the boy, and it was feared that he had fallen into the sea and drowned.

The Mayor, who lived more than two miles away, noticed that his Irish setter Harpo was running to the front door, and scratching, as though he wanted out.

The Inexplicably Bizarre

Since this was out of character for the dog, the Mayor followed the dog, to a crevice hidden by undergrowth, where the semi-conscious boy had crawled.

How did the dog, more than two miles away, know that Oscar was even missing? Let alone the fact that the dog knew where the young lad lay.

David Bellamy, who wrote the forward for the excellent Book *Psychic Animals* by Dennis Bardens, from which several of these condensed accounts have been taken writes: "That animals possess strange and unexplained powers has been generally recognized for thousands of years. That monuments have been erected, at various times and in different countries to animals and birds is further confirmation of the respect, and sometimes awe, in which they have been held. Many commemorate acts of sacrifice and bravery, of disasters averted or lives saved by the intervention of an animal."

A stone wolfhound on the walls of Antrim Castle, Ireland, is a memorial to a wolfhound that saved Lady Marion Clothworthy from a wolf attack in 1660 and also alerted the castle to a sneak onslaught by enemy forces.

On December 10, 1919, a dog was credited with saving 92 lives. The S.S. Ethie, a coastal steamer of 414 tons, was aground on Martin's point off Curling, Newfoundland, and breaking up in a violent storm, and heavy seas. The Captain was unable to fire a lifeline or launch its boats, and no member of the crew dared attempt to swim ashore.

A Newfoundland dog made the swim with a lifeline gripped in its teeth. As a result, all 92 passengers and crewmembers were pulled to safety on a boatswain's chair.

The late Professor J. Gaither Pratt, had no doubt that animals possess extra-sensory perception. He quoted one case in particular which is quite baffling.

A twelve-year-old boy had a pet pigeon, which had a permanent aluminum band on its leg with the identifying number 167. The boy was in a hospital some seventy-five miles from his home, when the pigeon appeared at the window, and began pecking on the glass. The nurse opened the window and in flew the boy's pet bird, identifiable by the number tag on its foot. How did the bird know where the boy was? How did he find his way there during a raging snowstorm? We may never know.

In Malmo, Sweden, in May of 1977, the two-year-old daughter of Leif Rongemo had crawled out onto a narrow ledge some four stories above the ground. Suddenly, Roy the Alsatian dog climbed out onto the same ledge, and seized the child's clothing in his jaws. Then, to the amazement of spectators below, he shuffled backwards, inch by inch, towards the window, and brought the child back into the room.

But should we be surprised? Why should not animals have mental, perceptual, or even psychic capacities in common with human beings?

There is in Edinburgh a monument to Bobby, a skye terrier that loved his master so much, that upon the master's death, he stayed at the grave for fourteen years, until he himself died.

A Spaniel owned by Jean Druisy, of Chateaulin, France, remained in the River Auline for six days and nights until its barking led neighbors to the body of its master - who had been murdered and thrown into the river.

Baron, a poodle given by French author Victor Hugo to the Marquis De Faletans and taken to Moscow, found its way back to Hugo's apartment in Paris, France, traveling 1500 miles in 3 months.

Loopy, a cat owned by W. Martin Ross, vanished from Ross' car in Medford, Oregon, on its first visit to that city and returned to Ross 3 days later at a house its owner had rented as a temporary residence - after the cat had vanished.

Mandor, a poodle owned by an Austrian customs official, nightly checked every sentry along the Austro-German frontier and barked to awaken any guard he found asleep. This dog policed an 18 mile stretch of border on his own, and was directly responsible for saving many lives, when he prevented several terrorist squads from entering the country.

The citizens of Freiburg in Germany raised a monument to a duck which saved thousands of lives during World War II, by warning the inhabitants of impending air raids by running up and down the streets with frantic quacking.

Dolphins have saved many human lives. Yet nobody has yet determined what makes dolphins so friendly toward people. There seems to be a rapport of some kind, perhaps even some type of mind-to-mind communication.

In one case, Bill Gomez was some distance from shore. He was nearing exhaustion, and feared he might drown. He began to sink beneath the surface of the ocean. His arms felt like lead. Then all of a sudden, a 10-foot dolphin got underneath Gomez pushing him to the surface. Then the dolphin started moving toward the shore. Gomez could feel the dolphin holding him afloat. In a few minutes, the dolphin dumped Gomez in the shallow water near the shore. This incredible friendly mammal had saved his life.

In another case, a seven-year-old girl from the Dominican Republic by the name of Becky was swept out to sea by a rip tide. A 12-foot dolphin came to her rescue. The dolphin came up between her legs, and the young girl literally rode the dolphin, like a horse, all the way to the beach, where she was deposited on the shore.

Why should a dolphin feel impelled to rescue from death someone of another species? Dolphins have to kill for food. Their huge mouths conceal a formidable array of razor-sharp teeth - yet they never bite a human being, not even if attacked. We may ask why a lawyer's wife swimming off a Florida beach in 1949, drawn out to sea by a series of mighty waves, should have been saved from death by a dolphin. As she took in water and was drowning, she found herself "pushed violently from behind, and I landed on the beach with my nose in the sand."

Eleven years later, off the same stretch of coast, a swimmer who was in difficulties found herself "guided" by a dolphin which nudged her away from the dangerous currents and into the safety of shallow water.

The Greeks also believed that a dolphin saved Apollo's son, Icadius, when he was shipwrecked near Lycia. The dolphin carried him on his back and landed him near Mount Parnassus, where in gratitude he founded a shrine.

Grecian coins show a man riding on a dolphin. In his *Natural History*, Pliny gives us the story of a dolphin that regularly ferried a boy across the broad arm of the sea to school at Puteoli. Pliny adds: "There is no end of examples of this kind."

In his presidential address to the Society for Psychical Research in 1962, Dr. Gilbert Murray asked: "Is there not some telepathy between animals and humans?"

V. L. Durov, an animal trainer, devised a series of experiments designed to answer this question.

He would give orders to dogs by telepathic means, making no sound of any kind, and in the majority of cases the dogs complied!

Dr. Bechterev, and several other scientists, confirmed Durov's conclusion that some type of "sixth sense" was operating. Professor Vasiliev, an associate of Dr. Bechterev, favored the view that telepathy is an instinct in animals, because they have no actual language. Clearly, if such a means of communication exist between animals, they could exist between animals and man also.

It is not easy to determine what is normal or abnormal in animal behavior. But it seems certain that animals, and dogs especially, have psychic powers. Josef Becker would agree to that. One night while Becker sat in a pub having a few beers, his dog became agitated, doing everything possible to attract attention, running round in circles, howling at his master, tugging at his clothes and trying to drag him from his stool. But Becker, intent on his drinking, put the dog out.

Somehow the dog got back into the inn by another entrance and began to tug frantically at Becker's leg.

Seeing that he would never enjoy his drink in peace, Becker left the pub by the front door. Exactly two minutes later, the inn collapsed on its occupants. Many were killed, and many more were injured. The stool that Becker had been seated on, only two minutes before, was totally crushed by a large beam. The dog had saved his master's life by some foreknowledge that the building was about to collapse. Had the dog's keen sense of hearing heard the beams cracking? Or did the dog have some kind of psychic power? By whatever power the dog clearly had a premonition of disaster.

The Watford Evening Echo in January 1971 published a story about a five-year-old Collie, Laddie, who accompanied his master every day to the slate quarry where he worked. This had been the dogs habit ever since he was a puppy, but one morning he refused to budge from the house. He ran about the house in a very erratic manner, howling, and pawing at his master. Robert Hays could not persuade the dog to come. So Hays for the first time, went to work alone. Just about noon there was an explosion at the quarry. Robert Hays was killed.

Dennis Bardens, who writes about this and similar cases in detail in his fine book *Psychic Animals* points out, "What explanation can there be other than that the dog somehow knew what was going to happen?"

Bardens relates another case in which a dog displayed not only heroism but also the intelligence to take an action that saved the lives of three little girls.

Mrs. Clarke, a platelayer's wife, was walking along the railway at Backworth in Northumberland, when a strange dog seized her coat and pulled her with such determination that at first she thought it was attacking her. Looking down, however, she noticed that, in gripping her coat, the dog had dropped a child's shoe from its mouth. The dog no sooner saw her looking down at the shoe than he picked it up again and rushed ahead. She followed the dog, which led her to three infants who were on the tracks.

It was not a minute too soon, for none of them was more than three years old, and all were oblivious of the danger, as a train approached. She was able to get the children off the track in just the nick of time. The dog had seen the danger and gone for help, with one of the children's shoes in its mouth.

The *Spectator* ran a story in 1804 concerning a man from Aberdeen who was crossing the frozen surface of the River Dee when the ice broke. The pieces around him were each insufficient to take his weight; as fast as he seized one, it broke in his hands. His plight was desperate. His dog watched this drama for about two minutes, and then disappeared and returned with a wooden pole, long enough for the man to rest upon two of the larger pieces, which he could hold on to until help came. Then the dog raced off to the nearest house and barked and yelped until help was forthcoming, leading the rescuers to the spot.

Cats also have been credited with saving many people. Dr. Ute Plemes, a German psychiatrist working at Giessen University, has collected over 800 cases of cats warning their owners of impending disaster. She concludes: "I have seen enough to convince me that animals do have a special psychic power to sense danger before it happens. We are fools to ignore them."

In one such case, a man was shaving himself in the German town of Magdeburg in 1944. While he was shaving, a cat's persistent meowing made him open the front door. On the step was a stray cat he had often stroked and fed. The cat pulled at his trousers as though demanding that he leave the house immediately. Very puzzled, the man hastily dried his face and followed the cat, which kept looking back at him to make sure that the man was really following.

When they had gone about 100 yards the cat stopped in its tracks. Overhead came the roar of Royal Air Force Lancaster bombers. One of the first bombs dropped, flattened to the ground, the house the man had just left.

The stray cat somehow had foreknowledge of this impending danger, and saved the man who had befriended her.

In another case, Betty White, who was 35 years old at the time in 1987, was working in her kitchen when an intruder came through the door and grabbed her. He knocked her to the floor, and with a knife at her throat began to tear at her clothing. She was in fear for her life.

Then without warning, Tabby the cat leaped onto the man's head, dug her claws into his face and eyes, and let out a long loud shrilling cry.

The man dropped the knife and ran from the house screaming, with both hands clutching the cat, which was still firmly "clamped" to his head.

Professor Gustav Wolf, a Basel psychiatrist, declared in 1914 "animals possessed some mysterious source of knowledge and awareness." More recently, Dr. Robert L. Morris, coordinator of research at the Psychical Research Foundation, Durham, North Carolina, declared: "Evidence that ESP is present in other species than our own is considerable."

There is, I am convinced, a mystical element in the close attachment between animals and humans. There is a case on record of a snake defending its master from attack, and the snake, as we all know, is thought of as the embodiment of menace.

John Carpenter kept a 10-foot boa constrictor in his garage as a pet. One Saturday afternoon John was feeding the snake. When an intruder, intent on robbing him, came out of nowhere and jumped on his back. Mr. Carpenter was dazed and lay on the cement floor bleeding. But all of a sudden the huge snake wrapped itself around the attacker. The snake had coiled itself around the man's arms and body. He could only scream for someone to help him.
Carpenter ran to a phone and called 911, and in a few minutes the police were there to arrest the culprit. Once they got him free of the snake.

The Forgotten Genius

Somewhere in the shadows of the early history looms the mysterious figure of Nathan B. Stubblefield. The son of a prosperous lawyer, he was born in Murray Kentucky, in 1860.

In 1892, years before Marconi could send and receive coded signals in a system that later became the telegraph, Nathan Stubblefield, at Murray, Kentucky, at points about two hundred feet apart had set up two boxes not connected in any way. Each box contained a telephone - and as Stubblefield and his son spoke to each other over the telephones, there voices could be clearly heard by the crowd who came to listen.

The Inexplicably Bizarre

"I have solved the problem of telephoning without wires through the earth. But I can also telephone without wires through space as well." Stubblefield said. "I can telephone without wires a mile or more now, and when the more powerful apparatus on which I am working is finished, combined with further developments, the distance will be unlimited!"

Everybody in Murray knew about Stubblefield's Black Box, which made the voices appear out of thin air. In 1892, he handed his friend Rainey T. Wells a box, and told him to walk away with it. "I walked a good distance when I heard HELLO RAINEY booming out of the receiver," reported Rainey. "There must be wires someplace!" I said. Wells continued to walk and all the while Stubblefield continued to talk to him.

The St. Louis Dispatch sent a reporter for a private demonstration. Stubblefield handed him a telephone, which was connected to a pair of steel rods about four feet long. He was told to take the phone anywhere he liked in the area, stick the rods into the earth, and put the receiver to his ear.

Just as Stubblefield had claimed, he could send messages through the air without wires. Nathan Stubblefield had conducted a public demonstration of wireless transmission of human voice for the first time in history.

In the newspaper article, the reporter told how he had gone about a mile from the inventor's house, put the rod into the earth, and, as he said, "I could hear every word that Stubblefield spoke!" Stubblefield told the newsman that he was merely using the electrical field, which permeated the earth, the water, and the Atmosphere. Stubblefield went on to predict that some day wireless transmission of speech would be used worldwide.

The citizens of that small town were affectionate toward Stubblefield, and years later in 1930 would erect a monument to him with the inscription: "THE FATHER OF RADIO."

In 1902, Stubblefield went to Washington, D.C., where he gave another public demonstration and amazed the scientists of his time with his wireless transmission and reception of the human voice. This was at a time when Marconi could only send and receive dot and dash code.

A telephone was installed on the steamship Bartholdi, and many scientists and dignitaries stationed themselves along the Virginia side of the Potomac. As the ship churned by, those on shore communicated with those aboard the vessel, the sound was clear and distinct, by merely sticking the customary iron rods in the ground and speaking into the telephones.

The Washington Evening Star said in headlines on May 21, 1902: - "First Practical Test Of Wireless Heard For half Mile."

Stubblefield continued to improve on his invention, and on January 15, 1903, perhaps in preparation for a patent application, his wife and children Bernard, Pattie, and Victoria signed the following affidavit:

> "In order to establish date of New Invention in Wireless Telephony of Nathan B. Stubblefield for any future needs that might arise technical description herewith provided of apparatus used. This day Nathan Stubblefield transmitted wireless telephone messages one hundred and twenty five yards without ground connection his latest development in wireless telephony. This affidavit is the first documented message transmitted by this system through sixty yards space. This message was transmitted at 8 o clock night of Jan. 15 by Bernard Stubblefield and received by Nathan B. Stubblefield the inventor and received again by the below signed as witness."

Since this device worked without a ground connection, it was totally dissimilar to his wireless telephone of 1902. Another affidavit, signed by Pattie and Victoria a year later, gave further details:

> "This is to certify that we the undersigned date above shown heard at a distance (roughly stepped) of six hundred feet harp music by Wireless Telephone, Nathan Stubblefield's secret invention where in no earth connection is used described as follows and understood by us. Circular coils of No. 28 magnet wire 26 ft. in diameter with forty convolutions with forty eight cell dry batteries connected in with coil and carbon ball transmitter, as transmitter of messages. Receiver as follows two coils wire seven feet in diameter containing 33 convolutions each. First coil office wire No. 20 second or top coil of No. 20 Magnet wire with two bell receivers. It is not understood by father or us whether it is by electromagnetic wave that this is done but well known that simply a primary current passes through coil and transmitter connected one to each distinct circuit or coil. Bernard B. Stubblefield transmitted music from coil just west of house, our home, to forked red oak tree on land east of our house, with its forks pointing north and south with poison ivy growing on its west side a snag of a tree with knot

near top rather on the south side. Given our hand this Sunday night Jan. 23 all with a view of establishing facts as they exist for the future interest of Nathan Stubblefield the inventor and our father who was with us in this test."

The final two affidavits came two weeks later, the first on February 4, read:

> "We the undersigned testify to the fact that this day a coil of No. 20 copper wire, the coil forty feet in diameter with 42 convolutions, with 48 cells of dry battery and a microphone [sic] transmitter was used in transmitting wireless telephone messages, conversation and harp music four hundred and twenty three yards from our residence with no sort of earth connection. A coil as receiver of 26 ft. in diameter of No. 28 magnet wire with 40 convolutions with a double pole receiver but no sort of earth connection. Other station lying westward in a woods from the home place located by a dogwood tree of small size known to us."

The second, also singed by the three children contains a note from Nathan at the end:

> "This is to certify that we the undersigned did this day receive wireless telephone messages, conversation and harp music four hundred and twenty-three yards distant from the transmitting station <u>without any sort of earth connection</u> by means of Nathan Stubblefield's new system of Wireless Telephony, claimed by him to be done through the Hertzian or electromagnetic wave process and practical for great distances, either stationary or portable.
> Note: The above are sons and daughters of mine who understand the technical features of my inventions."

Stubblefield went home with financiers begging him for contracts. Sometimes his neighbors saw him from a distance and reported seeing mysterious lights that lit up the whole area. Strange sounds could also be heard coming from the vicinity of his modest home.

Two weeks before his death, Stubblefield visited Mrs. Owens a close neighbor, and told her: "I've lived fifty years before my time. I've perfected now the greatest invention the world has ever known. I have now taken light from the air and earth as I did with sound. I want you to know that I can now make a whole hillside blossom with light.

Meanwhile his personal life deteriorated. Stubblefield had always been very secretive. He never permitted visitors on his property because he feared they might steal his inventions. He put all of his money into his electrical projects and experiments, and lived in abject poverty. He died of starvation in 1928. He was found dead in his crude shack, his equipment missing, his inventions gone, and his records scattered.

With the death of Stubblefield, this particular avenue of research died with him. The telephone using the magnetic field of the earth was never developed.

On the courthouse lawn at Murray, Kentucky, a stone memorial marks the place where he made history in 1892.

Time Warp

On a cold stormy day long ago, Sir Victor Goddard, somehow flew his biplane four years into the future!

In the year 1935, while a wing commander, Sir Victor Goddard flew his Hawker Hart biplane to Edinburgh, Scotland from his home base in Andover, England for a weekend visit.

On Sunday, before flying back, Goddard took a tour through an abandoned World War One airfield in Drem, near where he had been staying. It was now being used as farmland. What buildings that were left standing were in disrepair.

On Monday Goddard began his return trip. The weather was very bad, and Goddard had lost his way in the maze of clouds and wind. Goddard was flying in an open cockpit, with rain beating down in his face. In order to get back on course, he decided to descend below the clouds. At the lower altitude, he could pick out a landmark. He knew that an old, abandoned airfield named Drem was somewhere in the area. He dropped the biplane down through the clouds, and there only a few miles ahead, lay Drem.

As he approached the airfield, an unearthly eerie light suddenly illuminated the entire area. It looked as if the sun was shining brightly, even though the cloud cover completely covered the sun. He flew over the airfield at less than 100 feet altitude. To his astonishment, the old abandoned airfield appeared to have been rebuilt overnight. Bright yellow aircraft lined the smooth and painted runway. He could see airplane mechanics, all dressed in blue overalls. The buildings were all new and brightly painted.

The Inexplicably Bizarre

Goddard was flying at 150 miles per hour, and just some twenty or thirty feet above the ground as he passed directly over the airfield. However, none of the mechanics seemed to notice him or hear his plane. In an instant the bright eerie sunlight that had bathed the field was gone, and as he sped away from Drem, he was again engulfed by the storm. But now he now knew the course to fly, in order to get home.

After he returned home, he reported that he was surprised when no one on the ground bothered to look up as his plane roared by overhead. He also reported that the area was extremely active, with mechanics in blue overalls working on planes that were painted bright yellow.

The report he made was met with many a raised eyebrow, for you see, Drem was abandoned in 1934, the year that Goddard made his flight.

However, in 1938, with World War two growing closer, the British reopened it as a flying school. And it was not until 1938 that British training planes changed their color from silver to yellow and that mechanics for the Air Force began wearing blue coveralls, instead of brown.

There was no way he could have known that in the future RAF would change the colors of their trainers and that the mechanics would switch to blue overalls. Goddard finally concluded years later that he must have traveled through some kind of time warp in got a glimpse of the airfield four years into the future.

Two Men Who Could Fly

The medium Daniel Dunglas Home was observed to levitate numerous times over a period of some 40 years and was never discovered in any fraud. F. L. Burr, editor of the Hartford Times, gave the first account of his unusual ability:

> "Suddenly, without any expectation on the part of the company [or on Home's part - he was 19 years old and this was his first, involuntary experience of levitation] *Home was taken up in the air. I had hold of his hand at the time and I felt his feet - they were lifted a foot from the floor! He palpitated from head to foot with the contending emotions of joy and fear, which choked his*

utterances. Again and again he was taken from the floor, and the third time he was carried to the ceiling of the apartment [the Connecticut home of Ward Cheney, a silk manufacturer], *with which his hands and feet came into gentle contact".*

That was in 1852. After that, he learned to control his flights. He demonstrated his ability to fly before emperor Napoleon III, and Mark Twain.

He went to England, where he gave a command performance to Lord Adare, his cousin Capt. Charles Wynne, and the master of Lindsay, later Earl of Crawford and Balcarres, Lindsay told the story:

"I was sitting on December 16, 1868, in Lord Adare's rooms in Ashley Place, London, S.W., with Mr. Home and Lord Adare and a cousin of his. During the sitting, Mr. Home went into a trance, and in that state was carried out of the window in the room next to where we were, and was brought in at our window. The distance between the windows was about seven feet six inches, and there was not the slightest foothold between them, nor was there more than a 12 inch projection to each window, which served as a ledge to put flowers on. We heard the window in the next room lifted up, and almost immediately after we saw Home floating in the air outside the window. The moon was shining full into my room; my back was to the light, and I saw the shadow on the wall of the windowsill, and Home's feet about six inches above it. He remained in this position for a few minutes, then raised the window and glided into the room feet foremost and sat down.

Lord Adare then went into the next room to look at the window from which he had been carried. It was raised about 18 inches; and he expressed his wonder how Mr. Home had been taken through so narrow an aperture. Home said "I will show you," and then with his back to the window he leaned back and was shot out of the aperture head first, with the body rigid, and then returned quite quietly. The window was about 70 feet high from the ground."

No mechanical supports or arrangement of ropes has been suggested in this case, and would not cover the facts as described.

The Inexplicably Bizarre

Adare and Lindsay clearly agreed that Home's flew and was seen to fly in through the window and, later, to fly both in and out of the windows 70 feet above the ground. And when he came back into the room feet first, there was no supports of any kind that could be seen or felt.

As for Captain Wynne's terse statement, it seems clear that he understood himself to be describing something quite out of the ordinary - for he denied being the "victim of a hallucination or of humbug."

In 1871, Home was observed to levitate and fly by Sir William Crookes, an eminent scientist who later became present of the prestigious British Association for the Advancement of Science. His statement printed in the Quarterly Journal of Science:

"The phenomena I am prepared to attest are so extraordinary, and so directly oppose the most firmly rooted articles of scientific belief - amongst others, the ubiquity and invariable action of the force of gravitation - that, even now, on recalling the details of what I witnessed, there is an antagonism in my mind between reason, which pronounces it to be scientifically impossible, and the consciousness that my senses, both of touch and sight, are not lying witnesses."

Many scientists and debunkers attended Home's levitations, but none were ever able to prove that Home was a fraud.

In 1873, Home moved to the Mediterranean where he died on June 12, 1886 at the age of forty.

Daniel Home was not the only man who could "fly", or that practiced levitation. One of the most unusual Christian saints, and the most susceptible to levitation, was Joseph Desa. Canonized on July 16, 1767 by Pope Clement XIII as Saint Joseph of Cupertino, he was born on June 17, 1603 to poor parents in southern Italy. When he was 17, Joseph attempted to join the Friars Minor Conventuals, but was refused admittance because of his lack of education. As he grew older, Joseph practiced increasingly harsh austerities. He ate only vegetables sparingly and at infrequent intervals.

In his early twenties, he was admitted into a Franciscan friary near Cupertino.

On March 28, 1628, he was ordained a priest. In the monastery Joseph was able to devote himself to his austerities.

STRANGER THAN YOU CAN IMAGINE

On October 4, 1630, the town of Cupertino held a procession. Joseph was assisting in the procession when he suddenly soared into the sky, where he remained hovering over the crowd. This was the first of many flights, which soon earned him the nickname "The Flying Saint."

Joseph's life changed dramatically after this incident. His flights came continued and came with increasing frequency. For the fact was that Joseph could not contain himself. On hearing the names Jesus or Mary, or while praying, he would go into a dazed state and sour into the air, remaining there until a superior commanded him to come down.

On one occasion, Joseph had said Mass in a private chapel and had then withdrawn to a corner of the church to pray. Suddenly, and without warning, he rose into the air and uttering a sharp cry, flew to the altar, his body upright and his arms outstretched. Seeing him alight on the altar amid the burning candles, several nuns began to scream: "He will catch fire!" But Joseph's companion, brother Lodovico, who seemed to have had some familiarity with such sights, assured them that Joseph would not be burned. After a short time, Joseph gave another cry and flew back from the altar, this time in a kneeling position, in which he landed safely on the church floor.

Joseph's most famous flight occurred during a papal audience before Pope Urban VIII. The Pope himself witnessed this levitation! Joseph had gone to Rome, where it had been arranged for him to see Pope Urban. Joseph quickly rose into the air where he remained for some time. The pope remarked that he himself would testify to the truth of what he had seen. Only when the Minister General of the Order, who was part of the audience, ordered him down was Joseph able to return to the floor.

There now began a period when Joseph's levitation were so frequent as to become his normal behavior!

Once he rose into the air and flew some 20 yards to the high alter, where he hovered for 15 minutes or so. Then he flew back to the floor.

Sometimes Joseph's flights took place outdoors. One day he promptly flew to the top of an olive tree, where he remained for about half an hour.

At other times Joseph was able to lift others with him into the air! One time he seized a man by the hair, rose into the air with him, and remained there for about 15 minutes.

In 1645, the Spanish ambassador to the papal court visited Joseph, and testified under oath to the following:

The Inexplicably Bizarre

"He at once flew about a dozen paces over the heads of those present. Hovered over a statue of Mary Immaculate for some time, then returned to the floor, and we all were speechless and astonished."

Another eminent witness of Joseph's flights was Johann Friedrich, Duke of Brunswick who said, "the monk gave his customary cry and rose into the air in a kneeling position, where he stayed for some time."

On August 10, 1663, Joseph became ill with a fever, but the experience somehow filled him with joy. He experienced ecstasies and flights during his last mass, which was on The Feast of the Assumption.

Joseph died on September 18, 1663. His last words, before passing, and while he was raised about six inches off the floor were: "Oh! What chants, what sounds of music, what perfumes, what sights of paradise!"

Because of his many flights, St Joseph is the patron of those traveling by air.

A film was made about St. Joseph of Cupertino entitled *The Reluctant Saint* starring actor Maximilian Schell.

St. Joseph of Cupertino's story was also made into a children's book called *The Little Friar that flew,* written by Patricia Lee Gauch and published in 1980 by Peppercorn Publishers. His Biography *Saint Joseph of Copertino* by Fr. Angelo Pastrovicicchi O.M.C. is available from Tanner Books and Publishers, Inc. The ISBN is 0-89555-135-7.

The Creation of a Life Form?

Andrew Crosse was born on June 17, 1784. In 1836 he was living in Sumerset where he had his own laboratory in his home in the Quantock Hills. He was also a local representative in the British parliament. Crosse had decided to conduct some experiments on the formation and development of artificial crystals by means of prolonged exposure to weak electric currents. For two weeks he had been passing an electrical current through a chemical solution in an attempt to induce crystal formation.

On the 26th day of the experiment he saw what he described as "the perfect insect, standing erect on a few bristles, which formed its tail"

Did he accidentally hit upon an arrangement, which created life out of inorganic matter?

More creatures appeared and two days later they moved their legs. Over the next few weeks, hundreds more appeared. Crosse identified them as being part of genus *acarus*.

In a paper, which he wrote for the London Electrical Society in that same year of 1837, Andrew Crosse set this account of his experience:

> *"On the fourteenth day from the commencement of this experiment I observed through a small lens a few small whitish excrescence or nipples, projecting from about the middle of the electrified stone. On the eighteenth day these projections enlarged and struck out seven or eight filaments, each of them longer than the hemisphere on which they grew.*
>
> *On the twenty-sixth day these appearances assumed the form of a perfect insect standing erect on a few bristles, which formed its tail. Till this period I had no notion that these appearances were other than an incipient mineral formation. On the twenty-eighth day these little creatures moved their legs. I must say now that I was not a little astonished. After a few days they detached themselves from the stone and moved about at pleasure.*
>
> *In the course of a few weeks, about a hundred of them made their appearance on the stone. I examined them with a microscope and observed that the smaller ones appeared to have only six legs, the larger ones had eight. These insects are pronounced to be of the genus acarus, but there appears to be a difference of opinion whether they are a known species; some assert that they are not.*
>
> *I have never ventured an opinion on the cause of their birth, and for a very good reason - I was unable to form one. The simplest solution of the problem that occurred to me was that they arose from ova deposited by insects floating in the atmosphere and hatched by electric action. Still I could not imagine that an ovum could shoot out filaments, or that these filaments should become bristles, and moreover I could not detect, on the closest examination, the remains of a shell. I then imagined, as others have done, that they might originate in the water and consequently made a close examination of numbers of*

vessels filled with the same fluid: in none of these could I perceive a trace of an insect, nor could I see any in any other part of the room."

A local newspaper published an article about the "extraordinary experiment" and named the insects *Acarus Crossii.*

Crosse suspected that the mysterious 'insects' might be stowaways on his porous stone, so he dispensed with the stone entirely and conducted experiments with glass jars filled only with various acids - copper nitrate, copper sulfate and zinc sulfate. Again and again his little visitors appeared in the solutions. He found, so he wrote, that these *acari* appeared first beneath the surface of the acid, but after once emerging from their acid birthplace, they quickly died if dropped back into it.

Some of the people apparently gained the impression that Crosse had somehow "created" the insects or at least claimed to have done so. He received angry letters in which he was accused of many bad things. Some of them included death threats.

Learned men denounced both Crosse and his "insects" as nothing more than the product of airborne spores or of impurities in the fluids themselves.

Crosse had anticipated the critics there. He had already begun a new series of experiments designed to determine whether or not the things he saw were the result of electrical generation and nothing more. Crosse boiled his jars. He baked some of the apparatus in an oven. He filled the receivers over inverted mercury troughs with manufactured oxygen and super-heated his silicate solutions. Crosse took these precautions to ensure the sterility of the materials used. Yet, after he sent the current into his sealed, air-free jars, the mysterious *acari* appeared as before.

Other scientists tried to repeat the experiment. William Henry Weeks took extensive measures to assure a sealed environment for his experiment by placing it inside a bell jar. He obtained the same results. Meanwhile, in an atmosphere strongly impregnated with deadly chlorine, Crosse produced still more of the creatures.

Standing virtually alone, he could only repeat that he had told the truth, as some of his detractors could readily determine if they would conduct similar experiments. Suddenly the assault upon him died down, for he found a champion whom the critics dared not question. The great Michael Faraday!

Faraday reported to the Royal Institution that he, too, had experienced the development of these little creatures in the course of his experiments. In February 1837 many newspapers reported that Michael Faraday had also replicated Crosse's results.

Crosse made no claim of having created anything. He merely tried to report to his contemporaries what he had seen and the conditions under which these things had transpired. He wrote:

> "Each mineral speck enlarges and elongates vertically: so it does with the acarus. Then the mineral throws out whitish filaments: so does the acarus speck. So far it is difficult to detect the difference between the incipient mineral and the animal; but as these filaments become more definite in each, in the mineral they become rigid, shining and transparent six sided prisms; in the animal, they become soft and waving filaments and finally are endowed with life and motion!"

Later researchers, such as Henry Noad and Alfred Smee, were unable to replicate Crosse's results.

Although he spent his life in the quest for scientific truths, the sole contribution for which Andrew Crosse is remembered is the mystery over his "*acari,*" a mystery that is still unsolved today!

Living Fish in the Pharynx

In his *Manual Of Medical Jurisprudence For India,* Chevers has collected five cases in which death was caused by living fish entering the mouth and occluding the air passages. He has mentioned a case in which a large catfish jumped into the mouth of a Madras Bheestic. An operation on the esophagus was immediately commenced, but abandoned, when the man died.

In 1863 White mentions that the foregoing accident is not uncommon among the natives of India, who are in the habit of swimming with their mouths open.

There is a case reported in the *Medical Times and Gazette* of 1863 in which a fisherman, having both hands engaged in drawing a net, and seeing a sole fish about eight inches long trying to escape through the meshes of the net, seized it with his teeth. A sudden convulsive effort of the fish enabled it to enter the fisherman's throat, and he was asphyxiated before his boat reached the shore.

The Inexplicably Bizarre

There is another case reported in *Medical Curiosities*, by George M. Gould, and Walter I. Pyle, in which a man named Durand, who held a mullet between his teeth while baiting his hook. The fish, in the convulsive struggles of death, slipped down the throat, and because of the arrangement of its scales it could be pushed down but not up; asphyxiation ensued.

Stewart, in 1886, published a case of a native of Ceylon who was the victim of the most distressing symptoms from the impacting of a living fish in his throat.

The native had caught the fish, and in order to extract it, placed it's head between his teeth, holding the body with the left hand and the hook with the right. He had hardly extracted the hook, when the fish pricked his palm with his long and sharp dorsal fin, causing him suddenly to release his grasp on the fish and voluntarily open his mouth at the same time. The fish quickly bolted into his mouth, and, although he grasped the tail with his right hand, the fish found it's way into the esophagus.

The fins on fish are directed toward the tail, which admit entrance, but interfere with extraction.

In 1873, Maclauren reports the history of a young man who, after catching a fish, placed it between his teeth. The fish, three inches long, by a sudden movement, entered the pharynx. Immediately suffocation ensued.

In another case in 1878, reported in the *Medical Times and Gazette* in London, a man while swimming had partially swallowed a live fish. The fish was about three inches long. Futile attempts to extract the fish by forceps were made.

Examination showed that the fish had firmly grasped the patient's uvula, which it was induced to relinquish when it's head was seized by the forceps and pressed from side to side.

There is an interesting account of a native of India, who, while fishing in a stream, caught a flat-eel like fish sixteen inches long. After the fashion of his fellows he attempted to kill the eel by biting off its head. In the attempt the fish slipped into his gullet, and owing to it's sharp fins, could not be withdrawn. Even after death it could not be extracted, and the man was buried with it protruding from his mouth.

BIBLIOGRAPHY AND ACKNOWLEDGMENTS

Special appreciation for valuable help in the text research is extended to these organizations: Library of Congress; The New York Public Library; The Orlando Public Library; American Museum of Natural History; The Smithsonian Institution; New York Botanical Gardens; New York Zoological Society; The Brooklyn Museum; The Metropolitan Museum of Art; Cornell University; Rare Books Division; National Aeronautics and Space Administration; Mutual UFO Network; National Investigations Committee on Aerial Phenomena; Center for UFO Studies.

Abell, George, and Barry Singer. *Science and the Paranormal.* New York: Charles Scribner's Sons. 1983.
Agee, Doris. *Edger Cayce on ESP.* New York: Castle Books, 1969.
Allegro, J. M. *The Dead Sea Scrolls.* New York: Penguin Books, 1956.
Anderson, Joan Wester. *Where Angels Walk.* New York: Ballantine Books, 1992.
Annals of the Anatomical and Surgical Society, Brooklyn.
Arnold, Kenneth. "I Did See the Flying Disks!" *Fate.* 1948.
Atwater, P. M. H. *Beyond the Light.* New York: Avon Books, 1994.
Bach, Richard. *Nothing by Chance.* New York: William Morror & Company, 1969.
Bannister, Paul. *Strange Happenings,* Grosset & Dunlap. 1978.
Barrett, Sir William F. *On the Threshold of the Unseen.* New York, E.P. Dutton & Company. 1917.
Barton, G.A. *Archaeology and the Bible.* 1937.
Beck, Theodoric and John. *Elements of Medical Jurisprudence.* New York: Little & Company, 1851.

Begg, Paul. *Into Thin Air.* London: David & Charles, Ltd., 1979.

Berlitz, Charles and William L. Moore. *The Roswell Incident.* New York: Grossett & Dunlap, 1980.

Bloecher, Ted. *Report of the UFO Wave of 1947.* Tucson, AZ: Institute of Atmospheric Physics, 1967.

Boeche, Raymond. *Bentwaters—What Do We Know Now?* Symposium Proceedings, 1986.

Bond, Frederick. *The Gate Of Remembrance.* New York: E.P. Duton, 1933.

Bond, Frederick. "Boom Times on the Psychic Frontier." *Time*, 4 March 1974.

Booth, John. *Psychic Paradoxes.* Buffalo: Prometheus Books, 1986.

Bowen, Charles. *The Humanoids*. Chicago: Henry Regnery Company, 1969.

Brandon, Jim. *Weird America.* New York: E.P. Dutton, 1978

Branson, Brian. *Beyond Belief.* New York: Walker, 1974.

Bray, Arthur. The Center for UFO Studies.

Brookesmith, Peter. *The UFO Casebook.* London: Orbis, 1984.

Budge, E. A. Wallis. *The Egyptian Book of the Dead.* New York: Dover Publications. 1967.

Burrows, Miller. *The Dead Sea Scrolls.* New York: Viking Press, 1955.

Burton, Maurice. *The Sixth Sense of Animals.* Taplinger Publishing Co., 1973.

Butler, Brenda, Jenny Randles, and Dot Street. *Sky Crash.* Suffolk: Neville Spearman, 1984.

Canadian Journal of Medical Science, Toronto.

Cazeau, C. J., and Stuart D. Scott, Jr., *Exploring the Unknown.* New York: Plenum Press, 1979.

Chalker, William. The Australian Center for UFO Studies.

Chevers, N. *A Manual of Medical Jurisprudence for India.*
Christopher, Milbourne. *ESP, Seers, & Psychics.* New York: Thomas Crowell, 1970.
Churchill, Allen. *They Never Came Back.* New York, Doubleday & Company, Inc., 1960.
Corliss, William. *Ancient Man: A Handbook of Puzzling Artifacts.* The Sourcebook Project, 1978.
Corliss, William. *Strange Universe.* Glen Arm, Maryland: the Sourcebook Project, 1975.
Coss, Thurman L. *Secrets from the Caves.* New York: Abingdon Press, 1952.
Craig, Richard. *The Edge of Space.* Garden City, New Jersey: Doubleday & Company, Inc., 1968.
Daily Express. London, 7 November1967.
Daily Mirror. London, 18 August 1978.
Daily Telegraph. London, 20 July 1966; 6 August 1981.
Davies, A. Powell. *The Meaning of the Dead Sea Scrolls.* New York: Mentor Book, 1956.
De Bergerac, Savinien Cyrano. *Voyages to the Moon & the Sun.* London: Routledge, 1927.
Dickason, Fred. *Angels, Elect & Evil.* Chicago: Moody Press, 1975.
Dingwall, Eric John. *Some Human Oddities.* London: Home and Van Thal Ltd., 1947.
Dingwall, Eric John. *Very Peculiar People: Portrait Studies in the Queer, Abnormal, and the Uncanny.* London, New York: Rider, 1950.
Donnelly, Ignatius. *Atlantis: the Antediluvian World.* New York: Gramercy, 1985.
Downing, Berry. *The Bible and Flying Saucers.* New York: Avon Books, 1970.
Ducasse, C. J. *The Belief in Life After Death.* Springfield, IL: Charles C. Thomas, 1961.
Edsall, F.S. *The World of Psychic Phenomena.* New York: David McKay Company, Inc., 1958.

Edwards, Frank. *Strange People.* New York: Lyle Stuart, 1961.

Edwards, Frank. *Strange World.* New York: Lyle Stuart, 1964.

Edwards, Frank. *Strangest of All.* New York: Citadel Press, 1956.

Ehrenwald, Jan. *The ESP Experience.* Basic Books, 1978.

Emenegger, Robert. *UFO's Past, Present, and Future.* New York: Ballantine Books, 1974.

Evans, Hilary. "What Happened at Roswell?" *The Unexplained*, Vol. 9, issue 102. (London)

Evans Wentz, W. Y. *The Tibetan Book Of the Dead.* New York: Oxford University Press, 1960.

Everybody's Weekly, 11 December 1954.

Finegan, J. *Light From the Ancient Past.* 1946.

Flammarion, Camille. *The Unknown.* New York: Harper and Brothers. 1900.

Flammonde, Paris. *UFO's Exist!* New York: G.P. Putnam's Son's. 1976.

Fodor, Nandor. *Between Two Worlds.* New York: Parker Publishing Company, Inc., 1964.

Freud, Sigmund. *The Interpretation of Dreams.* The Modern Library, 1950.

Gaddis, Vincent. *The Strange World of Animals.* Cowles Book Co., 1970.

Gaddis, Vincent H. *Mysterious Fires and Lights.* New York: David McKey Company, Inc., 1967.

Garrett, Eileen. *Telepathy: In Search of a Lost Faculty.* New York: Helix Press, 1974.

Godwin, John. *Unsolved: the World of the Unknown.* New York: Doubleday & Company, Inc., 1976.

Goldman, Karen. *The Angel Book.* New York: Simon & Schuster, 1992.

Goldman, Karen. *Angel Voices.* New York: Simon & Schuster, 1993.

Good, Timothy. *Above Top Secret.* New York: William Morrow and Company, Inc., 1988.

Gould, George M. and Walter L. Pyle. *Medical Curiosities*. London: Hammond Publishing Ltd., 1982.

Green, Celia. *Out-Of-Body Experiences*. Institute of Psychophysical Research, 1968.

Greenhouse, Herbert R. *Premonitions: A Leap Into the Future*. New York: Bernard Geis, 1971.

Greenhouse, Herbert. *The Astral Journey*. New York: Avon Books, 1976.

Group, David. *Early Man and the Cosmos*. New York: Walker and Co., 1984.

Group, David. *The Evidence for the Bermuda Triangle*. London: Aquarian, 1984.

Hall, Angus. *Signs of Things to Come*. London: Aldus Books Ltd., 1975.

Hansel, C.E.M. *ESP and Parapsychology*. Prometheus Books, 1980.

Harrison, Michael. *Fire From Heaven: A Study of Spontaneous Combustion in Human Beings*. New York: Methuen, Inc., 1976.

Hayman, Leroy. *Thirteen Who Vanished*. New York: Julian Messner, 1979.

Hintze, Naomi, and Pratt J. G. *The Psychic Realm*. Random House, 1975.

Hitching, Francis. *The Mysterious World: An Atlas Of the Unexplained*. New York: Holt, Rinehart and Winston, 1978.

Hitching, Francis. *The Mysterious World*. Holt, Rinehart & Winston, 1979.

Holroyd, Stuart. *Dream Worlds*. Garden City, New Jersey: Doubleday & Company, Inc., 1976.

Holroyd, Stuart. *Minds Without Boundaries*. Garden City, New Jersey: Doubleday, 1976.

Hopkins, Budd. *Intruders*. New York: Random House, 1987.

Hopkins, Budd. *Missing Time*. New York: Random House, 1981.

Hugel, Friedrich von. *The Mystical Element of Religion*. London: J.M. Dent & Company, 1923.

Hynek, J. Allen. *"The UFO Update." Omni Magazine*, April 1984.

Hynek, J. Allen. *The UFO Experience.* Ballantine Books, 1974.

Johnson, Kendall. *The Living Aura.* New York: Hawthorn Books, 1975.

Kanigel, Robert. *Why Dream?* Science 85, December 1985.

Kastenbaum, Robert. *Between Life and Death.* New York: Springer Publishing, 1979.

Keyhoe, Donald. *Flying Saucers From Outer Space.* New York: Henry Holt, 1953.

Keyhoe, Donald. *Aliens from Outer Space.* London: Panther Books, 1975.

Koestler, Arthur. *The Roots of Coincidence.* New York: Random House, 1972.

Lair, Pierre-Aime. *Essai sur les Combustions Humaines.* Paris: Crapelet, 1800.

Le Poer Trench, Brinsley. *The Flying Saucer Story.* New York: Ace Books, 1966.

Lincoln, Jackson Steward. *The Dream in Primitive Cultures.* Johnson Reprint Corp., 1970.

Lipsie. *Miscellanea Curiosa Medico-Physica Acad.* 1670.

Lorenz, Konrad. *Studies in Animal and Human Behavior.* Harvard University Press, 1970-71.

Los Angeles Examiner, 23 May 1955.

Luce, J. V. *Lost Atlantis: New Light on an Old Legend*, New York: McGraw-Hill Books, 1969.

Mackenzie, Andrew. *The Unexplained.* London: Arthur Baker, 1966.

Mackenzie, Norman. *Dreams and Dreaming.* The Vanguard Press, Inc., 1965.

Manchester, Richard B. *Incredible Facts.* New York: Galahad Books, 1986.

Mardindale, Cyril C. *The Message of Fatima.* New York: Kennedy, 1950.

Memoirs of the Medical Society of London

Michel, Aime. *Flying Saucers and the Straight-Line Mystery.* New York: Criterion Books, 1958.

Michel, Aime. *The Truth About Flying Saucers.* New York: Criterion Books, 1956.

Mitchel, John, and Robert Rickard. *Phenomena: A Book of Wonders.* London: Thames & Hudson, 1977.

Moody, Raymond. *The Light Beyond.* New York: Bantam Books, 1988.

Moody, Raymond. *Reflections on Life After Death.* New York: Bantam Books, 1977.

Morse, Melvin, M.D. *Closer to the Light.* New York: Villard Books, 1990.

Muldoon, Sylvan. *Psychic Experiences of Famous People.* Chicago: The Aries Press, 1947.

Myers, Frederic. *Human Personality and Its Survival of Bodily Death.* Longmans, Green & Co., 1903.

Nash, Jay Robert. *Among the Missing.* New York: Simon and Schuster, 1978.

National Aeronautics and Space Administration

National Archives and Records Administration, Washington D.C.

National Security Agency

New York Times, 3 February 1934; 23 August 1946; 11 October 1946.

Noorbergen, Rene. *Secrets of Lost Races.* Indianapolis: Bobbs Merrill, 1977.

Ortzen, Len. *Strange Stories of UFO's.* New York: Taplinger Publishing Company, Inc., 1977.

Ostrander, Sheila and Lynn Schroeder. *Psychic Discoveries Behind the Iron Curtain.* New York: Bantam 1971.

Owen Robert Dale. *Footprints on the Boundary of Another World.* Philadelphia: J.B. Lippincott & Co., 1875.

Penfield, Wilder. *The Mystery Of the Mind.* Princeton University Press, 1975.

Pfeiffer, C.F. *The Biblical World.* Grand Rapids: Baker Book House, 1966.

Phillips, Perrott. *Out of This World: The Illustrated Library of the Bizarre and Extraordinary.* London: Phoebus Publishing Company, 1976.

Phillips, Phil. *Angels, Angels, Angels.* Lancaster PA: Starburst Publishers, 1995.

Reader's Digest Editors. *Strange Stories, Amazing Facts.* New York: The Reader's Digest Assoc., 1976.

Reynolds News, 15 June 1957.

Rhine, J. B. *Extra-Sensory Perception.* Bruce Humphries Publishers, 1964.

Ripley's Giant Book of Believe It or Not. New York: Warner Books. 1976.

Rogo, D. Scott. *Mind Beyond the Body.* New York: Penguin Books, 1978.

Sanders, N. K. *The Epic of Gilgamesh.* New York: Penguin Books, 1960.

Saunders, W. B. *Anomalies and Curiosities of Medicine.* 1896.

Schwartz, Stephan. *The Secret Vaults of Time.* New York: Grosset & Dunlap, 1978.

Scully, Frank. *Behind the Flying Saucers.* New York: Henry Holt & Company, 1950.

Seers, Stan. *UFO's: the Case for Scientific Myopia.* New York: Vantage Press, 1983.

Shepard, Leslie. *Encyclopedia of Occultism and Parapsychology.* Gale Research Co., 1978.

Sigstedt, Cyriel. *The Swedenborg Epic.* New York: Bookman Associates, 1952.

Steiger, Brad. *Mysteries of Time and Space.* Englewood Cliffs, New Jersey: Prentice-Hall, Inc., 1974.

Stemman Roy. *Mysteries of the Universe.* London: Aldus Books, 1980.

Story, Ronald. *The Encyclopedia of UFO's.* Garden City, New York: Dolphin Books, 1980.

Stuttman, H. S., *Mysteries of Mind Space and Time.* Westport Co: Orbis Publishing Co., 1992.

Sunday Dispatch. London, 28 March 1954; 7 November 1954.

Tanous, Alex. *Beyond Coincidence*. New York: Doubleday, 1976.

"The Randlesham Forest Mystery." *Flying Saucer Review*, June 1982.

The Reader's Digest Association, Inc., *Into the Unknown*, 1981.

The Reader's Digest Association, Inc., *The Worlds Last Mysteries*, 1978.

The Roswell Daily Record, 8 July 1947.

The Times. London, 25 November 1954.

"The Truth About Flying Saucers." *Look*, 17 June 1952.

Tyler, Kelsey. *There's an Angel on Your Shoulder*. New York: Berkley Books, 1994.

"UFO's and Project Blue Book." United States Air Force, Public Affairs Division.

Ullman, Montague, and Krippner. *Dream Telepathy*. Penguin Books, 1974.

United States Medical Investigator, Chicago.

U.S. News and World Report. March 1991.

Vallee, Jacques. *Forbidden Science*. New York: Marlowe & Company, 1996.

Vallee, Jacques. *The Invisible College*. New York: E.P. Dutton, 1976.

Vallee, Jacques. *Passport to Magonia*. Chicago: Henry Regnery Company, 1969.

Vallee, Jacques. *UFO's in Space: Anatomy of a Phenomenon*. Chicago: Henry Regnery, 1965.

Vaughan, Alan. *Incredible Coincidence*. New York: J.B. Lippincott Co., 1979.

Von Daniken, Eric. *Chariots of the Gods*. New York: Putnam's Son's, 1970.

Von Daniken, Eric. *Gods from Outer Space,* New York: Putnam's Son's, 1970.

"What You Can Believe About Flying Saucers." *Saturday Evening Post*, 30 April 1949.

Wilkins, Harold. *Flying Saucers on the Attack.* New York: Citadel Press, 1954.

Wilson, Colin. *Enigmas and Mysteries.* London: Aldus Books Limited, 1976.

Wilson, Edmund. *The Scrolls from the Dead Sea.* Oxford University Press, 1955.

Wilson, Ian. *The After Death Experience.* New York: William Morrow and Co., 1987.

Wolfman, Benjamin B. *Handbook of Parapsychology.* Van Nostrand Reinhold Co. 1977.

Yadin, Y. *The Message of the Dead Sea Scrolls.* New York: Simon & Schuster, 1957.

Zaleski, Carol. *Otherworld Journeys: Accounts of Near Death Experience in Medieval and Modern Times.* New York: Oxford University Press, 1987.

ABOUT THE AUTHOR:

Donald P. Coverdell, Th.D., received his doctorate in theology from the International Bible Institute & Seminary in 1985. Dr. Coverdell is a noted expert on Egyptian and Biblical records, and has written extensively about the Dead Sea Scrolls and the 4,000 year old Sumerian and Babylonian tablets. He has made a lifelong study of the paranormal, the extrasensory, along with the supernatural. He is the author of four books and numerous articles on these and related subjects. Dr. Coverdell has appeared on a number of radio and television programs dealing with these and other subjects.

From these and other activities, he has gained a rather large following, and enjoys some national name recognition. He currently lives in Orlando, Florida.